The Edge of Evolution

The Edge of Evolution

Animality, Inhumanity,
and Doctor Moreau

RONALD EDWARDS

OXFORD
UNIVERSITY PRESS

OXFORD
UNIVERSITY PRESS

Oxford University Press is a department of the University of Oxford. It furthers
the University's objective of excellence in research, scholarship, and education
by publishing worldwide.Oxford is a registered trade mark of Oxford University
Press in the UK and certain other countries.

Published in the United States of America by Oxford University Press
198 Madison Avenue, New York, NY 10016, United States of America.

Cataloging-in-Publication data is on file at the Library of Congress
ISBN 978–0–19–021209–4

9 8 7 6 5 4 3 2 1
Printed by Sheridan, USA

CONTENTS

PREFACE

I've taught many different university courses, first as a biology graduate student, then as a professor: freshmen and capstone, majors and non-majors, focused and general, biology and interdisciplinary. Pound for pound, it was the non-majors general biology classes that taught me the most about what people "out there" know, how they think the world works, and what order and type of information opens up the discussion. I loved these classes, because the students were unswervingly honest: screw up some ideas, or give an unfair or poorly conceived test, and you found out about it in no uncertain terms, and not months later in an evaluation, either. But if you show that you're exacting but fair, and bring the biology into their lives—not as spectacle, but as a genuine issue—then they show up early, arrive waving something they've looked up, stay late, make friends with one another, and study diligently.

They also taught me the most about the limitations of my course material: what parts were hand waving, or seemed like it, and what gaps were obvious to them although invisible to me and my colleagues. I made it my professional business to keep refining the presentation and the content for maximum clarity, to keep reorganizing the material for maximum impact, and to keep redesigning the testing methods for maximum reward for learning. I also tried many ways to open the course—what exercise or project should kick it off, and what to say. My operating principle, the most successful as it turned out, was brutal honesty to set up a real social contract:

> The administrators call this a "requirement." I do not actually know if they laugh about the tuition they're getting from you, but they do wear tassel loafers and have much nicer offices than anyone I know at this school. I can tell you, they are not my people, and I do not work for them. I work for one guy down the hall, the department head. I teach this class because I think it's important. I don't think the tuition is "extra" or a scam, and I'll tell you why, or I hope to show you why as we go along.

But I know I can't ask you to believe that, not right now. Professors tell you all kinds of things on the first day and unless something has changed, well, I don't remember a lot of my old profs really making good on it. I'm saying that you are right to ask, how is this worth my time? Why do I have to do this? I'm also saying that I can't answer it unless you meet me—let's say, a third of the way. Attend class, do some stuff, and see what makes sense—talk about what doesn't, ask some questions, and in a few weeks, see what you think. Really be here, though, don't wander in with your head somewhere else. I'll be doing the same, because although I do have a schedule of topics, there's some flexibility, and I'll always say it exactly as I think you, no one else, right here this term, will get the most out of it based on what's happened so far.

Here are some mechanics to help us both with this.

- I don't assign discussion points, subjective points, whatever you want to call them. Your grade comes right out of your scores. That's because I don't trust myself, or anyone, not to abuse those points for students I like. This way, you can raise questions, you can be wrong, you can disagree with me, or you can simply keep to yourself if you want, and there's not a thing I can do to you in terms of your grade. It keeps that safe.
- I don't take attendance or apply it to your grade. You're all grown-ups. I know most of you work, like I did in college. You have to decide how to trade off your time and your obligations. So you know, we'll have graded work almost every day, and I don't do "ten percent off" or make-ups. If something truly medical or outrageous is involved, it's pro-rated, and that's all. Miss one without one of those situations listed in your syllabus, and it's a zero—but the good news is, no extra points are coming off.

At about this point, the students are surprised, and I can see them thinking, "This guy might actually be all right." That's when I tell them something biologically amazing, not just a cool detail from a nature special, but a point they never dreamed of, which makes them think back on their experience and sense their own bodies differently. It could be any of a dozen things, from the blood pooling in their circulatory system because they've been sitting too long, to what in the world is actually in that cup of legal psychoactive drug I'm drinking in front of them and what it's doing to my brain, or anything else immediate, experiential, and familiar. Halfway through the explanation, I ask, "Do you want to know?" And the tigerish enthusiasm that responds lets me know, this term, this class, we're going to make it.

Author's Voice

I've written this book in the language I developed in these classes. It's still a professor voice, although I hope only the good parts: intellectual ruthlessness, attention to the listener's starting position, and the biologist's typical and possibly charming social shortcomings. But it's also a fellow learner's voice, ready to be surprised by what the other humans might say. I tell every classroom of students at the beginning of a course that they are not intellectual subordinates, that although they may well be informed and provoked in a good way, they don't have to agree with me in order to pass, or to make me feel important. I learned as a student myself—and carried it into my teaching—that instruction is not about rank, it's about showing you can add value and about building trust.

Here, I'd like to reduce the implicit authority of writing a book as far as I can. My own history—teacher, evolutionary biologist, reader of speculative fiction, political activist, animal care committee member, small-press game publisher, father and husband, and more—produce a certain lived expertise, perhaps a good one for this project. I entered my studies in an exciting period for evolutionary thinking, and in retrospective, had remarkable luck in meeting and working with some notable individuals. Instead of a single mentor and specialty, I learned techniques from the fragments of DNA—back when gel kits were built in the lab—to hands-on work with both living and dead animals, to the hectares of open fields full of unsympathetic plants and creatures who bite. I had also been lucky to be trained in a broad range of liberal arts, with an eye toward history and toward the multiple ways a problem or idea is expressed in a historical moment; science to me is never merely testing a hypothesis, it's a powerful debate in which this one study flares up briefly, and that debate neither came from nowhere nor exists in isolation from anywhere in society.

Two things stood out for me, beyond the technical science. The first was the role that we as scientists found, or made, toward those animals we held power over, and who would be so surprised as me to find myself in a position of authority toward other scientists as well—some pretty scary smart ones, too, with many thousands of animals in the picture. The second was teaching, which I love, the job that never goes sour or old, the wonderful blend of the familiar and reliable, combined with the new things I'd decided to bring to it this time and the new statements and thoughts a new room of students may bring. It's added up: history, evolutionary theory, animal care, and the demand and excitement of finding whether it makes sense in the meeting of minds. One might say I've been training all my life to be able to talk about this single novel.

I've written from that expertise, as how could I not, but my point about the author's voice stands: I do not assume the privilege to inflict the technical information upon a captive audience, or to use it to baffle others into silence. I do not

claim to be an objective eye; I have lots of positions to present and a few irrepressible opinions that I hope to acknowledge as such. You'll come to see my views on
such things as metaphysical reality, religious outlooks, God, human beings, the
mind, animal welfare and rights, and more. It would be amazing if you didn't, and
dishonest of me to slip them past as authoritative or obvious. I've tried to make
it clear whether I'm explaining someone else's view, representing current biological thought, or going off on my own. With any luck, we can find a unique contact
between my argument and your own lived expertise.

Almost everything about this project, then, flies in the face of the accepted
forms of writing science, history, and history of science. With respect to these
many colleagues, love you as I may, I am not writing for you. I haven't allied my text
with specific camps within evolutionary biology, or among the current schools of
ethics, or with one or another clump in the thickets of literary theory. I'm writing
instead to people as I've met them through my students, who are generally willing to learn once they're convinced the course design is a fair deal, and after some
time, discover a new depth of astonishment and curiosity. Therefore I'm using a
couple of pedagogical principles from those classroom as well: first, to put aside
the ideal of complete coverage, preferring instead the "pump" model of inspiring
students to do and learn things. I do not use the standard web of technical terms
for such things as natural selection or metabolism, but instead the same language
I've developed over the years in class, beginning with familiar and informal terms
to lay a foundation for learning technical ones, introducing an important point as
a secondary topic with a single interesting detail, to be presented more completely
later. The key is to maintain accuracy, such that simplification strikes home rather
than glosses anything over.

Similarly, the usual academic completism in referencing is not productive in
this context. I've chosen references as doors to further reading, much as with
handouts in a course, not as an exhaustive list of whatever anyone may ever have
thought about a given topic, with the only exceptions being direct attribution for
specific points.

The tone comes from my classroom as well, and although the grammar is probably considerably better in written form, it's also true that impolite words may
appear here and there.

Taking Things Apart

Here's what I say when I pull out an ominous-looking cardboard box. It's one of
the few times when I shift out of my ordinary sunny classroom demeanor.

> A word on specimens and dissections. The animals you'll be working
> with came from slaughterhouses, as in our culture we kill pigs in great

numbers every day, and some of them are pregnant. We'll be using late-term embryos, effectively functioning infants at the time of their death. Ordinarily they would be incinerated with the rest of their mothers' viscera, but some have made their way here, as a detour, and when we are done, they will be similarly incinerated. Understand me well: these animals were in our power; their entire existence was subject to our control. Our use of them here is a sub-routine of their deaths in that context. Those deaths are our responsibility, both as a culture and here in this classroom with you. That use, then, is a serious matter to me. There will be no failure to attend carefully to everything they can teach us, nor especially any disrespect to their bodies. I will react to that as I would if these were the bodies of human infants donated to science education. Is that understood?

I also talk about what dissection is for. I tell them, this isn't *pro forma*. It's not a scavenger hunt in which they check off a list of things, pointing to them in the animal as they go.

We're looking at something that can be seen and taught no other way—that the zillions of individually living cells don't make a body without what's called organ systems. They aren't engineered. They aren't "made." They're historical elaborations on moving stuff around the body in bulk, generating—effectively by accident—an interior environment, a place where the cells live, and as it's happened, in combination, the larger organization we call the "organism," the "individual," or "you." Your cells are alive, but you as such are not. "You" are, by definition, this maintained internal environment and the chemical and electronic interactions among these systems. You may think you're looking at a pig's guts, but your organ systems are the same as his or hers, and what they do is you, in a way that a single cell of yours could never be. You're here to learn those systems today: their parts, yes, but also what substance each one moves around, from where to where, and what happens to it.

In class, the students work in pairs, and my rule is that every animal will be dissected, but that no student is absolutely required to touch the body. Squeamish people have to find willing partners. I enjoy that rule quietly, because that day or the next, I'm accustomed to seeing the formerly reluctant students right in there, gloves on, pointing to this and that, exemplifying active learning and peer teaching. Nearly all of them do that.

I've written a book to dissect a book, for very similar reasons that I include animal dissections in class. It's more than simply taking it apart. It's understanding what it does, and experiencing—I hope—specific shock about it. I think *The*

Island of Doctor Moreau is an important book, maybe the most important I've read, and it just may be that I can lead this dissection right.

- Part I is about orienting toward the novel, or how one might approach it as a reader. All of the chapters in this part are historical in one way or another, but Chapter 1 is mainly about philosophical and emotional ways to categorize the words "human" and "animal," and serves as the introduction for the book as a whole. Chapter 2 is mainly about the political and intellectual framing of the issues, and Chapter 3 is mainly about narrative presentations regarding science.
- Part II is about research using nonhuman animals, with Chapter 4 focusing on pain and animal welfare, and Chapter 5 on experimental design and animal rights. I disclose a bit about my own history in these areas, and provide some thoughts on Moreau's ideas as they'd be tested today.
- Part III is about human and nonhuman identity, with Chapter 6 focusing on cognition and appearance, and Chapter 7 on species differences and social behavior. I focus on the changing views of the main characters of the novel and suggest that some of them aren't the named humans.
- Part IV is about the presumed humanity of social institutions, with Chapter 8 focusing on religion and Chapter 9 on different aspects of violence. Chapter 10 presents some conclusions and questions about the impact on science, morality, and our understanding of society if the term "human" carried no intrinsic meaning beyond a common species designator.

Given all the history and science, many details and ideas will be pulled in, but I've tried to keep them from taking over. When all is said and done, I really just want to talk about this book as a story and its possible importance to ourselves, today. For instance, my interest in H. G. Wells as a historical person starts and stops with his writing it. His interpretation of Huxley's *Evolution and Ethics* (1896) past that point, his essay "Human Evolution, An Artificial Process," his views toward the future of humanity, his back-and-forth views on eugenics, his range of political positions, his personal life, and most of his other writings have no place in my argument, and I'm not explaining or judging them. I'm completely ignoring his psychology and intentions; you won't find any "Wells means this" or "with this, Wells is saying."

Although this position borrows a bit from the famous deconstructionist concept of "the death of the author," I don't wade in those waters very much either. My reading is more pedestrian, oriented toward the fictional content: who the characters are, where they are, what they do, and what happens to them, much in the same language as people ordinarily discuss stories they like. I like this one a lot, and what's more, I think it's genuinely valuable. Instead of deconstructing the plot, I prefer to value its structure and to engage with what fictionally happens. I dissect it in order to appreciate how it works and to enjoy it even more, not to strew its parts around.

To know this story well, history matters a great deal, and I aim to explain its relevance to the point of near madness, but not to dismiss the content of the story itself by embedding it safely long ago and far away. I've chosen what to include partly to provide context for what in the world some particular phrase in the novel is about, but mainly to bring in historical points that jump into the reader's lap, as they are often surprising roots for the familiar issues of today. Classic historical completeness simply has to pay the price—instead of trying to get every historical event in there, I've tried instead simply not to be absurdly wrong about it. I've included some timelines to split the difference.

This was a lot harder to do when writing about evolutionary theory and specifically about Charles Darwin, who exerts a gravitational pull of fascination for many biologists. The hardest revisions always came from remembering that we already have many books on the subject and many biographies of the man, and that I was going off on a crazed tangent again, so wham, there went another unnecessary fifteen pages. Although plenty of technical information did stay, this isn't a textbook; I'm teaching insofar as I'm opening a door to possible interest, but not by providing a mini-encyclopedia.

Squishy Words

Some words simply haven't lent themselves to easy definition, so will take on subtly different meanings depending on the immediate topic. I hope to avoid confusion in each case through context.

Most obviously, "human" and "animal" are under direct criticism as terms—that's mainly what this book is about. Their use falls into these categories:

- As dichotomous components in an opposed pairing, recognizable as capitalized, as in the Man/Beast divide, or tagged with judgmental adjectives like "exalted Man, lowly Beast." In all cases, this usage is the target of my disagreement with it, and indicates a position or point of view that I do not share, and in these passages, to which I am giving voice in order to critique.
- As quotes or vocabulary from the novel, which often expresses confusion or dogmatism on the part of the narrating or speaking character. I've provided a summary of such usage in Chapter 6.
- As standard informal scientific terminology, in the clumsy arrangement of "human" and "nonhuman animal," which doesn't manage to capture the reality. In Chapter 7, I discuss the difficulties of the terms "human animal" and "nonwolf animal," with "wolf" being only a familiar species.

The Establishment, capitalized, was a mostly derogatory term during the nineteenth century, and the particular confluence of power it referred to was a real entity, so my usage usually applies specifically to that. However, it may creep in as a more general term, if the same phenomenon applies in a relevant way to a current topic.

The Progressive movement during the nineteenth century, or progressivism, is usually applied toward historical groups and people who used the word themselves, and I do not apply it to today's political spectrum, except insofar as specific groups have chosen to do so.

Charles Darwin's term "natural selection" has undergone both cultural and scientific changes since its coining over a century and a half ago. When referring to it as a historical term, I call it by its original name, "natural selection," but when discussing it as a biological process, I merely call it "selection," with the implication that nuances such as artificial, sexual, and others are present on a case-by-case basis.

ACKNOWLEDGMENTS

I am grateful for past influence and encouragement from my mentors and colleagues Gregory Foster, Dale Hinckley, Steven Proulx, Forbes Keaton, and Robert Klevan; Jay Schleusener, Robert J. Richards, Noel Swerdlow, Joseph Williams, Michael LaBarbera, and John Bolt; Andrew Zimmerman and Mathew V. Jones; Bruce Patterson, Larry Heaney, Julian Kerbis, and Bill Stanley; Karla Hahn, John Bertram, Mark Blumberg, and Pamela Austin; Richard Kiltie, Michael Miyamoto, Peter Feinsinger, Carmine Lanciani, David Evans, Brian McNab, Vasiliki Smocovitis, Jack Putz, Donald Dewsbury, and Robert Lacy; Michael Lacey, Cathy Langtimm, and Carlos Martinez del Rio; Philip Frank and Dustin Penn; Stanley Cohn, Elizabeth Leclair, Anthony Ippolito, and Rima Barkauskas.

In warm memory, I am also grateful to John Powers, Aron Moscona, Philip Hershkovitz, Howard Moltz, Frank Kinahan, John F. X. Eisenberg, Stephen Jay Gould, and David Hull.

It has been an honor to know and work with Philip Iannacone, Kristen Kenney, Jose Hernandez, Lisa Forman, William Tse, Ranna Rosenfeld, Marilyn Lamm, and Xiao-di Tan of the former Children's Memorial Research Center, now the Stanley Mann Institute of Research.

I am grateful to the many students in non-majors biology courses at the University of Florida from 1989 through 1998, at Valdosta State University from 1998 through 1999, and at DePaul University from 2000 through 2014, and that biology capstone class from spring 2012, of course.

I thank the Robert Louis Stevenson graduating class of 1983 for a well-timed motivational moment during 2013, and also my friends Tod Olson, Julie Stauffer, Maura Byrne, Max and Alanna Lazarowich, Keenan Farrell, and Sam Shenassa. Closest of all, and with gratitude and love, I am grateful to my wife Cecilia Friberg for her support and kindness through the writing of this book.

PART I

NOT MEANT TO KNOW

Unless *suffering* is the direct and immediate object of life, our existence must entirely fail of its aim. It is absurd to look upon the enormous amount of pain that abounds everywhere in the world, and originates in needs and necessities inseparable from life itself, as serving no purpose at all and the result of mere chance. Each separate misfortune, as it comes, seems, no doubt, to be something exceptional; but misfortune in general is the rule.

—Arthur Schopenhauer, "On the Suffering of the World,"
Studies in Pessimism,[1] p. 207

A woman is immobilized in straps and chains, subjected to surgical interventions for six weeks, without anesthetic. She is carefully kept alive, healing each time, then operated upon again. She used to be a puma, or in American English, a mountain lion, but the techniques are frighteningly effective. Even five weeks ago, an unprepared observer thought she was human, and now she is—as the surgeon says—the apex, possibly the crowning moment, of his techniques. He has altered her posture, her skin, and her fur; he has slit and reconnected muscles, broken and reconnected bone. He has worked extensively on her brain, using long needles through the back of her mouth. She screams every day. One day, the surgeon arrives and she is waiting for him with a new look in her eyes and a new sound to her voice—because finally, her chain has come loose from the wall. You decide which of the two is more human, in this moment.

Two men confront one another on the sands of a beach. Nearby, others are knee-deep in the water, committing several corpses to the waves, but these two stand apart. They know each other well, and each despises the other's past dishonesty. One is subtly distorted in his form, such that you'd think he's a person, but something about his powerful build and his expression hints at his origin, a combination of hyena and swine. He stands without threat, his hands at his sides, but full of mute challenge: "Well?" The other has a broken arm held in a sling; when he turns to face the other, they stare at one another, then he throws down the whip he's carrying and draws a pistol. He cries, "Salute! Bow down!" and the first man defies him. He shows his teeth—not human at all, the only overt sign—and says, "Who are you, that I should"— then the other shoots at him. One of these men has a deep ethical and social question to resolve, but the other, the one born from a human man and woman, is consumed by unthinking fear and anger, and by the urge to kill.

I want to tell you about two lectures and two books, each lecture as challenging now as it was in its day, and each book written as a response (Figure I.1).

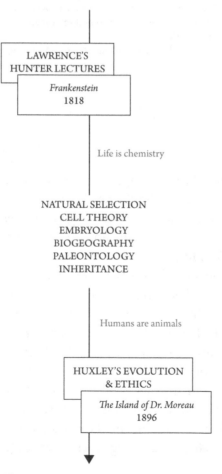

Figure I.1 Lectures and books.

The first lecture was by William Lawrence, daring to suggest that life, with humans in it, was a matter of physics and chemistry—that we are not *from* anything special. Mary Shelley, a friend of Lawrence, wrote the first book, titled *Frankenstein*. The second, over seventy years later, was by Thomas Huxley, daring to suggest that humans as animals had not evolved *to* anything special, socially or morally. H. G. Wells, a student of Huxley's, wrote the second book, titled *The Island of Doctor Moreau*.

Both novels matter a lot, and I'm writing to examine what they mean to us today, especially the second. I've pulled the scenes above from it. Are the characters I've described human? Does "being human" mean anything important? It's not only a good question, it is a powerful, unresolved question even all these years since, suitable for a book of its own.

Note

1. Arthur Schopenhauer, "On the Sufferings of the World," Studies in Pessimism, reprinted in Collected Essays of Arthur Schopenhauer, Wilder Publications LLC, 2008.

1

The Paw

In about my twentieth or twenty-first year of teaching university biology classes, my students were preparing for an exercise, and I was explaining a couple of terms. One of these was homology—roughly, the same organs in different creatures—and as an example, I mentioned the arrangement of bones at the end of the front leg of vertebrate tetrapods. The term "hand," I said, indicated one version of the same arrangement observed across all these creatures, and indeed ours is not much changed from its ancestral form.

Warming to my point, and encouraged by the little grunts or glances that I'm always looking for in class, I then truthfully reported that my wife—a veterinarian—and I had been changing our wiggling infant twins' diapers just the past night or so, and one of us had said to the other, "Hold his front paw."

One student's face distorted in surprise, and she impulsively asserted, "That's fucked *up*."

Professors differ regarding what to do at such moments. I typically encourage an informal classroom, and my amused shrug apparently settled whatever tension might have arisen, so we all went on to the next phase of the day's activity. But my little mental checklist received another mark: once again, I had seen the great divide in worldviews concerning humanity.

I'll grant you, the student may not have personally and specifically represented that divide. It could be that she knew that the term "animal" or any slang-free reference to it is demeaning, and reacted to that, without necessarily buying into it. If so, it merely kicks the can down the road, though, to ask why that encoding exists, for her to tap into it that strongly. That would be the better question anyway—why culturally and symbolically the distinction is so available—rather than considering it in terms of one person's psychology.

There's a little biological detail to think about, too, which is that the human hand is evolutionarily less changed from its ancestral vertebrate structure, say, that of a salamander, than what we typically call a paw, say for a cat or dog. If we wanted to get derogatory with terms like this, it would make more sense to call one of their paws a "hand."

It's a divide. Humans are animals, or they aren't, with any of the latter's weasel phrases like "came from," "but," and "not just." This divide is no joke, not only because it's deep, but because it's invisible. You can't use education and profession as a guide. I've encountered it among other academics, including biologists, and even among evolutionary biologists and anthropologists. Nor is religious commitment an indicator. I've found that it might crop up among religious people of various persuasions, but it might not—that is, their positions may be the same as those of anyone else. And as my student demonstrated, one's position on one side of it or another may not be acknowledged to oneself: even she did not know how strong her reaction was until it arrived.

How about you? Imagine, it's diaper-changing time. The little children, boy and girl, are about six months old, able to sit up and scoot a little, but not yet

---

crawling easily. Right now, they're lying comfortably with their arms and legs waving about, smiling and cooing. One parent is a biology professor specializing in evolutionary theory, and the other is a veterinary specialist, but right now, they're enjoying what to anyone else is a noisome job, wiping and diapering side by side, exchanging smiles with the kids.

One child, let's say the boy, waves his arms energetically, creating minor issues with the nearest parent's task. She says to the other parent, "Hold his front paw." Both of them laugh, and to them it's funny mainly because it's true. The joke is not based on the misapplication of the term, but on the opposite—its complete aptness.

So, I ask you: Is the love shown by the parents in this situation diminished by knowing their son's hand is a type of paw? How about their hopes for his future: Do they have less hopes, or different ones, from anyone else? Are they prone to treat him differently from other parents in terms of safety, discipline, health, or anything similar? Less formally, "What *exactly* do you think is so fucked up? What value does this distinction you're making add, for anyone or anything?"

Taking Exception

This distinction has a name: human exceptionalism, also called anthropocentrism. That's the viewpoint that asserts that members of the species *Homo sapiens* are distinctive from all other living creatures, past and present, in ways that transcend ordinary or typical species differences. Or, as so many accounts state it, that we are not only unique in the sense of being a particular species, which can be said about any gopher or fungus or whatever, but in the sense of a *kind* of difference that merits special comment and consequences: "uniquely unique." It is also—crucially—*above* the alternative (Figure 1.1).

Human exceptionalism takes many forms, but it always includes two distinct components, each with a special implication.

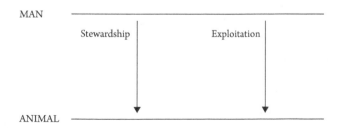

Figure 1.1 Exceptionalism.

- To assume or perceive a special inherent quality in humans, not merely in degree of difference from other creatures, but in kind: the uniquely unique feature. It could be cognitive, anatomical, spiritual, philosophical, cultural, or whatever, but it is *something*. Even if you can't name it, you don't have to, because you know it's there.
  - Also to include a consequential special ethical capability, by which humans judge their own and others' actions, and adjust their own actions individually and as groups, that other creatures do not have.
- To infer or expect a distinctive role that humans are expected to take on, which they may either be doing already, or should start doing as soon as possible. This role is a sociological or political one, whether among fellow humans, directed toward non-humans, directed toward the larger economics and ecology, or all of them at once. It isn't merely a preference or desire; you see it as a genuine and indisputable directive, inherent in human existence, which is literally immoral to violate or ignore.
  - Also, to infer or expect a special fate, either awaiting our species or already in progress. Exactly what this is ranges all over the map and could be a book in itself.

The mantle of superiority is suspiciously vague. Any metaphysical outlook can serve—specific or vague, religious or secular. You can frame the unique quality in any way you like. You can use the language of God and the Bible, the Deep Green concept of Gaia, intuitive spiritualism, the racist distortions of Social Darwinism—any system of classification, even without explicit mystical content, as long as it includes the concepts of higher and lower, or even merely a subtle and unarticulated emphasis concerning superiority. That's how more than a few evolutionary biologists and physical anthropologists can uncritically maintain what they call "weak" exceptionalism. For them, it's all right to say a hand is a paw, as long as you frame it so that a hand is *better* than a paw, and that the processes resulting in a hand were either a special sort of evolution or a special achievement of evolution, not the boring kind that all those other creatures got or did.

The obligatory role is probably the real motor at work, because it is always explicit. It's also completely flexible in terms of what one is directed to do, so that both sides of a controversy may share the presumption of special human privilege and responsibility. When that happens, it's an intellectual and political disaster: the discourse goes around and around, neither side ever gaining ground, because to attack the other's argument at its root would be to undermine one's own.

I'll use the issue of how we, people, should act toward nonhumans or, more generally, toward "nature." It's a fixed feature of exceptionalism that we simply must adopt some special role toward the other creatures because of the separate and elevated status, as shown by the two levels in Figure 1.2. The arrows in

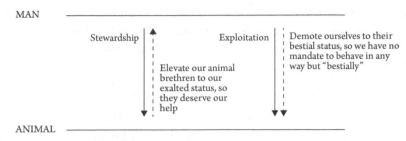

Figure 1.2 Exceptionalism, stewardship, exploitation.

Figure 1.2 illustrate that both stewardship and exploitation can rely upon this shared framework, such that they compete for the status of having this special, directed, and obligatory goal. You can also play tricks with how you phrase the human superiority in ways that give both sides something apparently fundamentally opposite to argue about. That's what the dotted lines are.

See how sneaky that is? Each statement looks as if it's willing to equate humans and nonhumans, and as if it's opposed to the other statement, but the *categories* of superior Man and inferior Beast are as fixed as ever. And even better, now the good-of-the-animals stewards and the good-of-humanity exploiters can butt heads all day long, each with a designated enemy, so the discourse can dance up and down those dotted lines. One side can even call out the others' exceptionalism and thereby keep its own obscured. The whole construct is intellectually empty, logically supporting nothing but available to justify anything, and so very attractive.

Human exceptionalism never, ever asks, "All right then, what *is* a human being, and how does it relate to the various other living things we see?" These are great questions—sadly, left behind in history, only partly addressed.

Science and Fiction

The exceptionalism question is all bound up with the origins of biological science, or the origins of a particular phase of it. Scientific thinking is as old as human culture, but modern, institutionalized science and its specific framework of ideas took their shape during the nineteenth century. The first big issue for biology was to make it a part of chemistry, which also came with making chemistry a part of physics. In *Distilling Knowledge*, Bruce Moran marks the transition for chemistry in the early 1800s, when it shed *superadded* forces, those of magic and metaphysics. This transition was dramatically evident between the eminent English surgeon John Abernethy, a key member of the Royal College of Surgeons, and his former student, William Lawrence, who was elected to the College in 1812 at the age of thirty. Abernethy considered the essentials of chemistry, magnetism, and electricity to be evidence of superadded forces, whereas Lawrence argued the more radical position, that these very things were wholly physical phenomena like anything else—and further, that being alive did not rely on a superadded quality.[1]

Lawrence had an ideal platform to present this view: the Hunter lecture series, which he began in 1816 and published as he went, with a summary volume in 1819 titled *Lectures on Physiology, Zoology, and the Natural History of Man.* The third lecture in the series and the introduction to the published volume include the clearest possible argument then or since:

> Life, using the word in its popular and general sense, which at the same time is the only rational and intelligible one, is merely the active state of the animal structure. It includes the notions of sensation, motion, and those ordinary attributes of living beings which are obvious to common observation. It denotes what is apparent to our senses; and cannot be applied to the offspring of metaphysical subtlety, or immaterial abstractions—without obscuring and confusing what is otherwise clear and intelligible. (William Lawrence, *Lectures on Physiology, Zoology, and the Natural Rights of Man,* Lecture II, p. 52)[2]

And,

> To talk of life as independent of an animal body—to speak of function without reference to an appropriate organ—is absurd. It is in opposition to the evidence of our senses and our rational faculties: it is looking for an effect without a cause. We might as well reasonably expect daylight while the sun is below the horizon. What should we think of abstracting elasticity, cohesion, gravity, and bestowing on them a separate existence from the bodies in which those properties are seen? (Ibid., p. 53)

Lawrence had struck a blow in the current culture wars about materialism and vitalism, terms with specific meanings at that time. Abernethy's vitalism held that living things contained or included some special essence that powered and literally animated them, making them alive and imparting certain properties; this essence was variously described, but it always included an expression of intent or what today we might call spirituality. Lawrence's materialism, on the other hand, meant physical properties affecting one another in reality, literally what you and I today call physics and chemistry. To explain life materially, you asked about and examined physical properties, more or less as if you were studying an engine. Lawrence had exposed the fact that many scientists in different fields were investigating life materially, and he provided a framework, even a name, for talking to one another about it.

The word *biology* in English dates from these writings and from this exact point. Although its precise author is obscure (Lawrence cited a German naturalist, G. R. Treviranus), credit for its adoption and its strict material basis lies with Lawrence, and his meaning requires careful attention: that life is a physical phenomenon, without the need for superadded forces or irreducible phenomena to explain or to study it. In a word, *biology* means studying life as a subset of chemistry, itself a subset of physics. Lawrence urged his audience to seek out every imaginable physical understanding of all organisms' diversity and functions they could conceive, and to understand them as physics and chemistry operating in the real world.[3]

He was surgically precise in his introduction to the *Lectures*, in direct reply to Abernethy, explaining that even if living things were special and consequential in some spiritual way, this way is not manifest in their structure, processes, and dynamics, which Lawrence called the "economy" of an organism. In other words, he did not attack anyone's belief in a spiritual existence or purpose, but rather he said something infinitely more powerful: that such belief would have to putter along on faith alone, without using the existence of living beings as a constant body of evidence.

From the very start of Lawrence's lectures, the establishment church, medical community, and educational curriculum recoiled in horror. English churchmen, politicians, and scholars perceived material thinking as a terrifying strike at morals, social stability, and their positions of status. They put much effort into articulating and supporting vitalism throughout several opinion-making publications like the *Quarterly Review*, including direct attacks on specific scientists and their investigations. As scholars were supposed to represent only the most well-regarded, upstanding virtues as judged by the privileged, such attacks were career destroying and amounted to thought policing (what today we call "control of the narrative"). Even as the lectures proceeded, Abernethy attacked Lawrence as both unpatriotic, for referencing foreign scientists, and immoral—even as conspiring to destroy morality. Lawrence responded defiantly in the collected volume, to say the least, and the infuriated Abernethy brought the full power of his privilege upon his former student to suspend him from his position, effectively threatening his social and professional destruction. In 1822 the Court of Chancery drew upon ancient laws to declare his book blasphemous and to rescind its copyright, which perhaps backfired as a cultural strategy, as it instantly went into multiple cheap printings and sold like gangbusters.

Lawrence was forced to repudiate his views in order to save his career, but in terms of raw science, he was the winner. From that point forward, scientific studies of organisms, and also of everything else, became more and more committed to the concept of *physical cause* as the sole topic of interest. This principle is so fundamental to our modern experience of science that it's hard to remember that previous forms did not share it. To claim that life is itself a physical phenomenon, such that analyzing its properties and causes need not consider any presumed

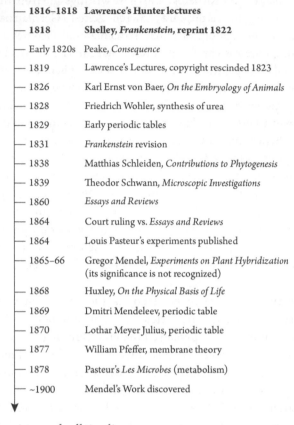

1816–1818	Lawrence's Hunter lectures
1818	**Shelley, *Frankenstein*, reprint 1822**
Early 1820s	Peake, *Consequence*
1819	Lawrence's Lectures, copyright rescinded 1823
1826	Karl Ernst von Baer, *On the Embryology of Animals*
1828	Friedrich Wohler, synthesis of urea
1829	Early periodic tables
1831	*Frankenstein* revision
1838	Matthias Schleiden, *Contributions to Phytogenesis*
1839	Theodor Schwann, *Microscopic Investigations*
1860	*Essays and Reviews*
1864	Court ruling vs. *Essays and Reviews*
1864	Louis Pasteur's experiments published
1865–66	Gregor Mendel, *Experiments on Plant Hybridization* (its significance is not recognized)
1868	Huxley, *On the Physical Basis of Life*
1869	Dmitri Mendeleev, periodic table
1870	Lothar Meyer Julius, periodic table
1877	William Pfeffer, membrane theory
1878	Pasteur's *Les Microbes* (metabolism)
~1900	Mendel's Work discovered

Figure 1.3 Chemistry and cell timeline.

purpose for life, or any nonphysical "vitalizing" property, is more than a mere "step" of scientific history. It's a game changer.

In the following decades, two concepts were developed: the periodic table of elements and cell theory, which together form nothing but the most exquisite confirmation of what Lawrence had suggested and lie at the heart of biology as we know it (Figure 1.3).

I'd like to focus on the chemistry for a moment. In the early 1800s, no one realized yet how few elements existed, how easily organized their interactions are, or how chemically unified all of known life is. It's one thing that Lawrence was right; it's another how simple the mechanisms are.

The chemistry is so easy you won't believe it. At the time of this writing, we know of 118 elements, but to understand the chemistry of life, you only have to know about five of them: carbon, hydrogen, oxygen, nitrogen, and phosphorus. Others are involved, but these five make the organic macromolecules (Figure 1.4).

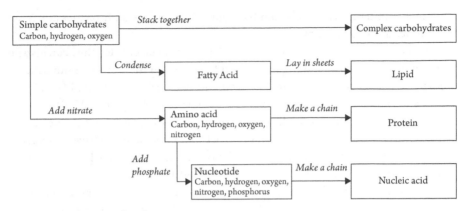

Figure 1.4 Macromolecules.

Carbon, hydrogen, and oxygen make two (relatively) little molecules called glucose and fructose. They're the carbohydrates—specifically, the simple ones. Every other named carbohydrate is merely a tinker-toy combination of these two, so those are called complex. Sucrose, table sugar, is fructose + glucose. Starch is another, merely bigger. Other words for "carbohydrate," exact synonyms, are sugar and saccharide.

In biology, the two simplest carbohydrates are the source of bodily energy for nearly every living thing on the planet, through the mechanisms of metabolism. More complex carbohydrates are either ways to store this energy prior to its use, or structural features of bodies, such as cellulose in plants, packed tight to become wood in some of them.

If you chemically condense a carbohydrate, so that it's smaller and (in energy terms) richer, it's a fatty acid. A bunch of fatty acids stuck together are a lipid; lipids get more complicated depending on what else is stuck to them, but the relevant point is that all cell membranes, on the whole planet, are based on the same lipid arrangement. Lipids are also used to store energy prior to use, such as fat in animals, and a few of them, the steroids, are hormones.

Before going on, imagine this: almost every cell on the planet has the same lipid membrane and the same metabolic activity, based on those three elements. That's right. Observed differences in those things are merely add-ons.

If you add nitrogen, technically nitrate, to a simple carbohydrate, you get an amino acid. There are only twenty of them, but any can hook up to any other in a chain structure—that's what makes a protein. Proteins are therefore the most diverse biochemical structures, effectively an infinite number of possible orders and lengths in a chain, each one subtly different in its reactions. Proteins are what make cells different from one another, through structures embedded in the membrane, through managing what gets in and out, and through whatever gets made in the cell and squirted out. Famous structural proteins include hemoglobin, pigment, and keratin, the stuff your hair and nails are made of. Enzymes and most hormones are also proteins.

If you add a phosphate to an amino acid, you get a nucleotide. They can also be combined into chains, making nucleic acids, including the universal inheritance molecules called ribonucleic acid (RNA) and deoxyribonucleic acid (DNA), as well as the famous metabolism molecule, adenosine triphosphate (ATP).

Living things indeed look differently than nonliving ones, and they do things that nonliving ones don't. They have structure, active and reactive properties, energy acquisition and storage (metabolism), and inheritance, when those macromolecules are organized in a certain way, called a cell.

That single tricky thought required decades of work throughout the nineteenth century by Matthias Schleiden, Rudolf Virchow, Theodor Schwann, Louis Pasteur, and August Weisman, culminating in the idea that cells are alive, formalized into a set of points called cell theory. This idea includes the possibly disturbing insight that creatures are heaps of cooperating cloned cells, organized as tissues, organs, and organ systems.

This is why biology teachers today seem obsessed with the interior of cells, right out of the gate. When Lawrence delivered his lectures, the outlook he presented so clearly was only that—a way to look at life. But the research he inspired eventually produced a core conclusion that cells, previously observed but not thought of in this way at all, were not only living things, but also the smallest individually alive things, or "units of life." You probably had to memorize that at some point, but maybe you missed the point: a body has no life of its *own*, but is an organization of living bodily cells constantly reproducing and dying, with its own properties. You as a person, with your name, your sense of existence, and your identity, are best understood as an organized property of tiny living things. What we call being alive, as people, is a matter of maintaining the relevant organization, for a while anyway, and what we call reproduction, as offspring or children, is a matter of some of the cells persisting in a new organization.

There are a million more details, as any biology major can tell you, but not one overrides what I've just explained: that the chemistry remains brutishly simple, no matter how complicated and gaudy the cells and their larger organizations may be. All the rest of biochemistry is about putting together and taking apart the macromolecules, and managing water using salt concentrations. Biological diversity—different creatures living in different ways—is a matter of add-ons and elaborations, defining the groups we call domains and kingdoms.[4]

Let's keep going with that "unit of life" idea. Because the cell is the smallest unit of life, therefore none of its components (membranes, chromosomes, etc.) are considered alive. Think about that! *Living creatures are made wholly of nonliving substances.* There is no "living stuff" in the universe, because being alive refers only to a certain category of how *some* nonliving stuff *sometimes* interacts. When you had to memorize that diagram full of cytoplasm and endoplasmic reticula, you were getting Lawrence channeled to you. It means that just because you are a *living* thing, you do not cease to be a *thing*.

I use the "you" with plenty of justification. Lawrence had not spared humanity the depth of his insights, far from it—fully half of the lectures' content is dedicated to human functions and diversity, described in the same terms and standards of analysis as any other organism.[5] The response included extensive, ongoing, and influential writing for the following six decades. The insight and its implications was definitely not lost on Friedrich Nietzsche:

> Let us beware of saying there are laws in nature. There are only necessi-
> ties: there is nobody who commands, nobody who obeys, nobody who
> trespasses. Once you know that there are no purposes, you know there is
> no accident; for it is only beside a world of purposes that the word "acci-
> dent" has meaning. Let us beware of saying death is opposed to life. The
> living is merely a type of what is dead, and a very rare type. (Nietzsche,
> *The Gay Science*, p. 168)[6]

Before these debates and writings began, however, someone else got there first, at the most relevant, challenging level. One of Lawrence's friends and an attendee of the lectures was Mary Wollstonecraft Godwin, newly Mary Shelley by mar-riage, a young scholar who not only perceived his points but immediately drama-tized them in her first novel in 1818: *Frankenstein: or, The Modern Prometheus*. But it wasn't the novel you may be thinking of.

The Created Man

The popular word "Frankenstein" is so entrenched as to defy meaningful reha-bilitation, even for those who know that the name refers to the scientist rather than his creation. It's identified with irresponsible research applications that transgress certain lines, promising woe to innocents, to the perpetrators, and to the creations themselves, who are some combination of malevolence and misery. It's what I call the "don't meddle" story, to be discussed further in Chapter 3.

The most-often read version of Shelley's novel was published in 1831, adding much talk of God, tagging the created man as soulless and self-loathing, and tag-ging Victor Frankenstein as a blasphemer punished for his transgression, as well as Shelley's introduction reinforcing all these points.

The original 1818 publication of *Frankenstein* isn't this story at all. The English scholar Marilyn Butler has provided great, wonderful service in calling atten-tion to the original, which saw only tiny print runs, and its republication usually includes her essay "Frankenstein and Radical Science." She outlines the direct link between the story and Lawrence's essays, the development of the alternate, vitalist message during the 1820s, and the pressure brought against Shelley to force the compromised edition.[7]

What's the difference between the two versions? In the 1818 version, the story centers on the twin facts that Victor Frankenstein is not a sorcerer who toys with forbidden forces, but a research engineer who works strictly with scientific chemistry; and that his creation is not a failure, or deficient in any way, but is instead a successfully created person. Frankenstein's methods are not arcane occult deviations from science, but rather an utterly physical extension of existing science. It forces the reader to cope with the fact that human beings are chemical, physical phenomena.

Crucially, Butler shows that the content in the sections cut from the original corresponds exactly to the material that the influential and hostile reviewer George D'Oyley in the *Quarterly Review* had insisted be cut from Lawrence's book. It was a cultural counterblow that combined these cuts, additional text to include the concept of Nature's or God's vivifying quality and tie it to an intrinsic morality, and Shelley's new introduction, which downplayed the novel's origin into a "ghost story" anecdote.

I've used the original text in class, and people struggle to read it fairly. Part of the problem is the movies' influence. I have to use short readings, in-class exercises, discussion groups, and focused questions for students to read the words *de novo* and finally to grasp that the creation is not a stitched-up corpse, that Victor does not use magic or even electricity, that he is not interested in being God, and especially, that his project is fully successful. They have to consider that Victor's judgmental language about the "demon," the "fiend," the "hideous wretch," should not be mistaken for the physical descriptions, none of which are quite as bad as the invective.

The difficulty continues with the plot events, in throwing aside the whole "the creation is rudimentary and goes berserk" model, and eventually realizing that the created man's capacity for evil is fully human, not inhuman or subhuman. In this, I have to help the students avoid looking for symbolic meaning or repeating what they think they know, and instead to focus strictly on characters' actions and responses to one another. When one of them says, "But it's a monster! It kills people!" I have them look again at whom the created man kills, and why. Then they realize that he is not a rampaging berserker, that this is not flailing-around killing, but rather that he *murders* people—exactly the way real people murder, motivated in ways a real person can understand and judge to be morally wrong, because a real person did it.

Victor's personal arc is marked by repeated reversals of his mood and goals, in the first half shown by his intense drive to create a human life flipping to his rejection and effective denial that he did it. In the second half of his story, he piles on such reversals faster and faster, until it's hard to see there's a Victor at all, instead of the next rebound from whatever he was just doing. His only act of thoughtful rather than impulsive, driven agency is to abort the woman he's creating in defiance of his agreement with the man. The created man's personal arc is extremely different and absolutely linear, each step marked by trauma and the development of more personal agency, from near-helpless infancy through suffering childhood, vengeful adolescence (which includes murdering William and indirectly Justine), and finally the decision to look his creator in the eye and demand recognition and the chance for a complete life, a promise he successfully gains. When that promise is betrayed, he quite precisely kills Victor's best friend and his fiancée, and indirectly his father, forcing Victor to be closest to none other but him and effectively to resign his fate into his, the created man's, hands (Figure 1.5).

The story is so grim and headlong because it's all about murder, really: the two early in the created man's life, to which Victor becomes his *post hoc* accomplice, and the two or effectively three committed by the created man in vengeance for the destruction of his mate. The rest of the story concerns their mutual flight into the Arctic, where Victor dies and the created man takes the body into the wilderness, presumably to die at the Pole.

The one moment at which the created man has the most wide-open decision regarding what to do with himself, and at which Victor has the most focused, non-rebounding opportunity to act, is when Victor aborts the created woman just prior to her completion.

Why does Victor do it? It's not a matter of the created man's wicked deeds to date—Victor actually gets that part, that they reveal him to be more human rather than less, and he knows very well that his own silence played the key part in Justine's hanging. He even agrees to the proposal of creating the woman, in exchange for the creation retiring to the wilderness—until he backs out at the last moment and destroys her. Again, why?

The text is explicit in two places. Regarding one of them, right before Victor violently dismembers her body, I agree with Butler's assessment that his mass of fears about the possible future of such a couple do not correspond to his priorities at any other point in the story and are best interpreted as excuses. I think more content is found in the other passage during the earlier confrontation, as he considers the created man's request and experiences two completely different feelings at once:

> His words had a strange effect upon me. I compassionated him and sometimes felt a wish to console him; but when I looked upon him, when I saw the filthy mass that moved and talked, my heart sickened and my feelings were altered to those of horror and hatred. I tried to stifle these sensations; I thought, that as I could not sympathize with him, I had no right to withhold from him the small portion of happiness which was yet in my power to bestow. (Mary Shelley, *Frankenstein: or, The Modern Prometheus*, p. 121)[8]

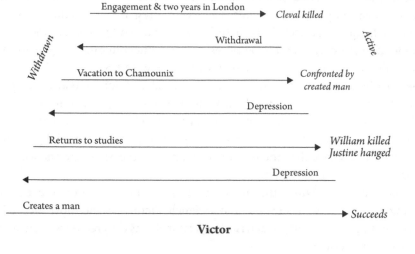

Victor

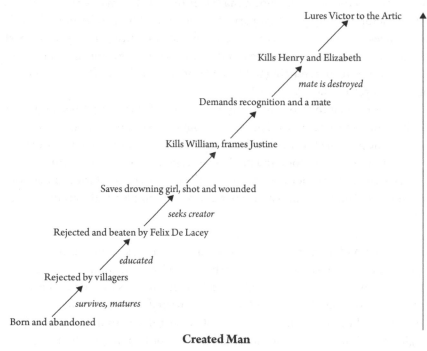

Created Man

Figure 1.5 The plot of *Frankenstein*.

Victor's invective toward his creation is usually repetitive, but at this moment it's unique and quite clear: he describes the created man as *physically* a "filthy mass that moved and talked." It's opposed in this passage to a different feeling of some empathy and responsibility, which at the moment he acts upon in agreeing to the proposed task. Ultimately, however, he acts upon the prior feeling of pure revulsion: he will not afford his creation even the status to be judged morally, or to consider his position or interests in any sympathetic way. In Victor's single moment of considered decision, he deems the created man *not alive*, and therefore out of bounds of human consideration of any kind.

That's the part that hits hard, way too hard for the Anglican political establishment of the early 1800s, and probably too hard for easy reading today. The created man *is* a person *and* a filthy mass, because—*that's what we, the real people, are.* Abernethy had denounced Lawrence with all the incoherent rage an ego-stung advocate of the exceptional Man could muster. This younger, junior scientist was deliberately, villainously seeking to destroy morality, he was in league with foreign agents (French ones! good God!), and he was effectively a psychotic intellectual terrorist. Why? Because Lawrence had frightened him with the utterly logical suggestion that aliveness is a matter of chemical operations, that real-world, real-life behavior is an organic function of the mind, and that although one may believe in a mystical soul, such a thing cannot be observed as the seat of morality, or as the seat of anything we see people really and actually do. This fright was not merely philosophical, but was politically associated with every bit of status and privilege that people like Abernethy enjoyed.

In the fiction, that is why Victor twice recoils from his creation, not because he's not alive, but because he *is*. When he begins even for a moment to consider the created man human, briefly to think of *his* emotions, Victor instantly retreats into hatred—because that humanity too evidently exists without miraculous mystery. He cannot bring himself to see that just because the created man is a *thing*, that does not prohibit taking him seriously as a *living* thing.

I don't expect you to agree with this reading—everyone knows the "monster" is hideous—even Wikipedia says so. Everyone knows it is "soulless," that it "should not exist," that Victor's constant and strict avoidance of personal pronouns (he, him) indicates what the book "means." But I submit that my reading is consistent with the original text and that the "soulless" reading is not.

Teaching by hammering students with some assertion doesn't work, nor is it my goal to force people to agree, or to pretend to. Here's what I did with them instead.

Format: 5 typed double-space pages with no extra spacing at any point, using standard margins. Include a cover page with your name, the title, the class information, and the date. Answer all four questions, in this order. Do not bullet or number them, and do not include any introductory or concluding paragraphs.

Identify a function or feature in Frankenstein's creation that must rely on some processes familiar to our own physiology, either because Frankenstein simply copied it or because he found a way to simulate it. (Example: we know the creation eats. There are others.) Explain how that function or feature works in our bodies.

Imagine a thinking creature whose body's physiology involves less physical substances than ours—perhaps mainly sunlight, gases, and connectors to conduct energy like electricity or light. Upon examining us and the human body to the extent that you have studied in this class, it pronounces a human to be "a filthy mass that moves and talks." Defend your honor and that of your species against this insult! Or conversely, agree with it. Either way, explain your position.

Is Frankenstein's creation human? Explain your answer in detail. You may have to divide the term into different meanings and address them separately.

You are a skilled cinematographer and film editor. Choose any scene in the novel and describe how you would design and create it in cinematic terms to achieve the maximum and accurate thematic impact you think it deserves.

It's quite wonderful to read twenty-five entirely different, thoughtful, slightly traumatized papers answering these questions, from students who've been studying macromolecules, cell structure, metabolism, anatomy, and physiology throughout the term. Few students consider literature and science class compatible, and even the science fiction and fantasy fans aren't accustomed to radical social science fiction: informed, confrontational, with no nostalgic looking back, and no idealistic looking forward.

Science Fiction

The nineteenth century rolled on and biology was becoming itself, as the modern forms of biodiversity, evolution, ecology, development, and genetics formed right along with cell theory. In some ways, the larger culture was recapitulating Victor Frankenstein's moment of reaction at the creation's birth: half in a state of excitement and half in a defensive crouch, as the scientific development of the question "What in the world is a human person?" was under way.

The question's focus would change, however. Whether humans are chemicals may be a profound issue, but it's abstract in terms of personal experience, ethical crisis, and setting policies. The ensuing debates and the issues around which entire political and cultural movements would coalesce were more visceral and personally jarring: whether humans were animals, specifically apes, and how humans were to treat nonhuman animals in a variety of contexts, especially scientific research.

The early context for these debates was the voluminous, anonymously published *Vestiges of the Natural History of Creation* from its first edition in 1844 through its twelfth by 1884. The final edition revealed its author to be Robert Chambers, who had died in 1871. This book was really the groundbreaker. Shocking in its presumption of material causes and human's place in them, it also managed to downplay the radicalism of the idea, pitching such causes as part of a grand design, and therefore more palatable to the English-speaking mainstream. Everyone read it, it spawned a reactive secondary literature, its anonymity lent it both a certain naughtiness and immunity from authoritative backlash, and its long run with multiple editions generated a culture of seeing "what's in it this time."

During the *Vestiges'* forty-year run, not a single work of natural history or other biologically related work can be properly considered without checking the publication date for its latest version, and its hottest topic was the historical and constant transformation of living things. Crucially, it subtly transformed Jean-Baptiste Lamarck's *evolution*, then associated with revolution, the bloody one in France, into *development*, a benign and ordered process. Even with that softening of content, it popularized the discussion and provided a common background and vocabulary for the larger culture's opinionated, interested debate about it—including the status of human beings as biological entities. Some of that material was hard core even by today's standards and might fairly be regarded as Lawrence's revenge:

> Less clear ideas have been entertained on the mental constitution of animals. The very nature of this constitution is not as yet generally known or held as ascertained. There is, indeed, a notion of old standing, that the mind is in some way connected with the brain; but the metaphysicians insist that it is, in reality, known only by its acts or effects, and they accordingly present the subject in a form which is unlike any other kind of science, for it does not so much as pretend to have nature for its basis. There is a general disinclination to regard mind in connexion [*sic*] with organization, from a fear that this must needs interfere with the cherished religious doctrine of the spirit of man, and lower him to the level of the brutes. A distinction is therefore drawn between our mental manifestations and those of lower animals, the latter being comprehended under the term instinct, while ours are collectively described as mind, mind being again a received synonym with soul, the immortal part of man. There is here a strange system of confusion and error, which it is most

imprudent to regard as essential to religion, since candid investigations of nature show its untenableness. There is, in reality, nothing to prevent our regarding man as specially endowed with an immortal spirit, at the same time that his ordinary mental manifestations are looked upon as simple phenomena resulting from organization, those of the lower animals being phenomena absolutely the same in character, though developed within much narrower limits. (Robert Chambers, *Vestiges of the Natural History of Creation*, chapter 17, pp. 325–326)[9]

By the 1850s, this discussion could no longer be controlled by clergymen and their academic allies, and younger scientists were publishing their ideas about it openly. Charles Darwin's *The Origin of Species* was one of these, first published in 1859, proceeding in six editions through 1872. Contrary to popular misconception, this book offered little regarding humans, but it was the watershed for both the theoretical underpinnings in our understanding of living things' diversity and the social clout to publish such things openly. One reason for that was the role of Thomas Henry Huxley, who decisively defended Darwin's theory of natural selection and the associated issues of human origins against all comers, fellow naturalists and theologians alike. The *Origin* gained its immediate social and intellectual fame at least as much from Huxley's spirited defense of evolution against Bishop Samuel Wilberforce in a post-seminar discussion at Oxford as from its own content, which is—between you and me—scientifically brilliant but a little bit bland when it comes to Man, Nature, God, and so on.

Speaking of nature, Huxley was one of its forces all by himself. Unlike Darwin and many of the other notable intellectuals in this period, he was raised not only middle class but also poor, and rather than groomed through privileged schooling, was apprenticed to harsh, grubby medical work in the 1830s, in the depths of Britain's worst depression. Self-taught in many topics from languages to philosophy to theology, he won top national awards in anatomy and physiology at the age of twenty; at the unbelievably young age of twenty-six, he became a Fellow of the Royal Society and was soon elected to its Council. Huxley was a walking challenge to the Victorian social hierarchy, curiously placed in its corridors of power, practically a time traveler from the next generation, free of the implicit indoctrination of both church and privilege, and he feared literally no one in debate or policy discussion.

Huxley was one of the few of this time who engaged directly with the question of humans as animals. His friend Charles Darwin was another, studying complex nonhuman behavior and writing of thought as a "secretion of the brain," and their respective work carried on the direct line of intellectual descent from Lawrence. Huxley's lecture "Evidence as to Man's Place in Nature" in 1863 provided the first really public, evidence-based demonstration that human beings were properly considered a type of animal—specifically, an ape. He also helped develop the link between cells and organisms in his lecture "On the Physical Basis of Life," in 1868.

By this time, the other major issue centered in human exceptionalism had developed dramatically as well: the general treatment of animals, with a special emphasis on the subjects of medical research. Huxley's prominent position in education and research policy placed him right in the center of the controversies concerning the Anti-Vivisection Bill in 1875, to be discussed in Chapters 2 and 4.

The early 1800s were a defining intellectual moment for biology as we know it today, and the 1890s were similar in terms of social positioning: the birth following the earlier conception, perhaps. Today's academic degrees, job descriptions, scientific organizations with their peer-reviewed journals, and other defining social elements of modern biology all took their form at this time. Intellectually, the different avenues of inquiry were tagged as a unit, and even if most of it was marked "to be figured out in detail later," these markings were now acknowledged. One of those unknown areas, now designated as such and recognized as crucial, was inheritance, and that's why Mendel's overlooked work was uncovered—it wasn't because someone happened to trip over it.

Evolution as an idea and, with it, human identity as a physical and biological phenomenon had suffered along the way. Early in the nineteenth century, it was primarily associated with radical politics and was considered shady, even subversive. Later, as a more mainstream term, it shook out into two distorted forms. The one favored by most intellectuals, deemed tolerable by the more powerful churches, and folded into scientific education, was a lovely confection of progress and the ascension of human to a current exalted position, and helpfully suggested what to do next. The other, made most famous by its vigorous articulation by industrial magnates in the United States and soon to take hold on the popular and cultural associations with the words "natural selection," "survival of the fittest," "evolution," and "Darwin," was a grubby justification of ruthless commerce and imperial colonial policy marked by considerable racism, even more disconnected from scientific content.

In 1893, in the Romanes lecture series at Oxford University, Huxley boldly defied both of these distortions, in the last lecture and publication of his life, *Evolution and Ethics*, which was published in 1895 with an additional *Prolegomena*. It was the Lawrence-style lecture of its day, entirely counter to the comfortable placement of scientific knowledge in the corridors of power and privilege. Huxley had not lost sight of the ideas that life and humanity are a *thing*, and that this thing happens to be an *animal*. How, he asked, is this to be reconciled with questions of morality and social purpose?

> Hence the pressing interest of the question, to what extent modern progress in natural knowledge, and more especially the general outcome of that progress in the doctrine of evolution, is competent to help us in the

great work of helping one another? (Thomas Huxley, *Evolution and Ethics*, p. 79 [1895 edn.]; p. 137 [1989 edn.])[10]

Here is my entirely unauthorized extraction, which I hope is not so unskilled as to draw much fire from the philosophers and historians of science.

Huxley broke with the tradition of viewing evolution as a march of ever more complexity and ever refined achievement. Instead, he describes a breaking down, or simplifying outcome of many processes that approach a more cyclical effect in the long term. It's not a genuine cycle so much as a frequent tendency toward destruction and loss, an effective simplifying of ecologies, which then serve as a new baseline, creating an illusion of cyclicity. In this, Huxley was right on target, although the history of mass extinctions would not be investigated seriously until over half a century later.

As if that were not enough, he introduced the unsettling consequence, the "baleful product of evolution," as he calls it, of suffering—raw pain. And in complete disagreement with just about everyone who was currently identified with the questions of evolution and humanity, he said that suffering is not alleviated by evolution, but made worse. Bluntly, life *hurts*. It is painful and unfair; there is no aspect of existence which imposes justice, harmony, or compensation. Human history did not escape from the bestial struggle into the rarefied halls of human society. It has nothing to do with anything triumphant, transcendent, or indeed, anything good.

There's a lot of Schophenhauer in there, especially the stark confrontation of the personal experience with the objective, and the simple denial of deep metaphysical understanding from material or, as we see it, real observation and thought. Huxley precisely dismisses chance as the operative factor, describing our history and our current rotten experience as the inevitable outcome of physical causes in succession. He refers to an "eternal immutable principle" of individual effort, later "the instinct of unlimited self-assertion," which is almost identically to the Will described by Schopenhauer, but goes the famously harsh philosopher one better, in removing metaphysical content from his description and landing us, humans, in an utterly physically determined condition.[11]

Huxley cited modern, political humanity as the state in which suffering has reached its most acute form. Evolution, natural selection itself, didn't make a happy animal, merely a very social one. We rather brutally minded creatures now find ourselves in a new social environment, in which our behaviors mainly accomplish abuse and oppression. Far from being an exalted or successful or otherwise perfected being, the human is a generally confused and bloodstained sufferer, with newfound ethics bashed around by long-established avenues of behavior.

For his successful progress, throughout the savage state, man has largely
been indebted to those qualities which he shares with the ape and the
tiger; his exceptional physical organization; his cunning, his sociability,
his curiosity, and his imitativeness; his ruthless and ferocious destruc-
tiveness when his anger is roused by opposition.

But, in proportion as men have passed from anarchy to social organization, and in proportion as civilization has grown in worth [*he clarifies
this later as "ethical" rather than civilized*], these deeply ingrained service-
able qualities have become defects. . . . they decline to suit his conve-
nience; and the unwelcome intrusion of these boon companions of his
hot youth into the ranged existence of civilized life adds pains and griefs,
innumerable and immeasurably great, to those which the cosmic process
brings on the mere animal. (*Evolution and Ethics*, pp. 51–52 [1895 edn.];
pp. 109–110 [1989 edn.])

They used to be ordinary, effective behaviors, but now they generate oppression, evils, crimes, and confusions. Bluntly, human life is especially miserable
because evolution has landed us in a situation we aren't good at. He specifically
denies that historical human societies represent stages of social or political development, identifying all recorded human history as in the same boat, which by
implication flew in the face of current British claims to reside at the apex of sociopolitical development.

Despite its complete reversal from contemporary narratives, this idea is dated
in some ways. It posits a wholly human ethics, however nascent and recent,
opposed to a wholly animal past, missing the complex social interactions of many
nonhuman species. It hints at Edenism, the idea that prehuman or nonhuman
existence was "better matched" to its environment, and therefore happier, or at
least it can be read with that underlying implication. The harshest modern reading of evolutionary theory suggests that selection is never about happiness, or that
the technical term "fitness" carries no implication of matching or fitting into any
appropriate conformation. But this view doesn't disagree with Huxley so much as
specify that even he didn't go far enough. Updating his essay to modern understanding actually makes his primary points even stronger.

His primary targets were well chosen. He attacked the claim to evolutionary
justification made by two groups: activists for reform, who over-idealize humanity's alleged "rise from the beasts" and expect miracles of justice to appear yesterday; and industrialists and imperialists, who exploit our alleged "brother to
the beast" imagery to justify their bullying and exploitation. He tossed out any
suggestion that society itself—specifically any distribution of privilege and
misery—has evolved or is evolving toward an ideal state. Natural selection has

not favored virtue over villainy, or villainy over virtue, however one may define those terms. In this, he accurately demolished current social appropriations of the words *evolution, natural selection,* and *Darwinian,* progressive and cynical alike. Human social effort to date, nice or nasty, is simply not the issue—what he sees is the injustice that prevails and our evident difficulty with, as he puts it so well, helping one another. What ethics have developed remain largely potential, in application.

> That man, as a 'political animal,' is susceptible to a vast amount of improvement, by education, by instruction, and by the application of his intelligence to the adaption of the conditions of life to his higher needs, I entertain not the slightest doubt. But, so long as he remains liable to error, intellectual or moral; so long as he is compelled to be perpetually on guard against the cosmic forces, whose ends are not his ends, without and within himself; so long as he is haunted by inexpugnable memories and hopeless aspirations; as long as the recognition of his intellectual limitations forces him to acknowledge his incapacity to penetrate the mystery of existence; the prospect of untroubled happiness, or of a state which can, even remotely, deserve the title of perfection, appears to me to be as misleading an illusion as was ever dangled before the eyes of poor humanity. And there have been many of them. [from the end of the *Prolegomena*] (*Evolution and Ethics,* pp. 44–45 [1895 edn.]; pp. 102–103 [1989 edn.])

Intellectually, this concept is a *tour de force* of beauty and horror, as it rightly recovers biology from its appropriation by all political efforts. Human exceptionalism is denied both its self-congratulatory rise from beast to angel and also its wallowing in exploitation, racism, and war with the excuse of being mere beasts.

He also broke with the metaphysical or spiritual model of human goodness, which at this time was generally comfortable with the idea that religion represents the highest and most fulfilling mode of human contemplation. Instead, he identifies the primary value in religious practice in the ascetic doctrines from some Hellenic philosophy, Buddhism, Hinduism, and some elements of Christianity—to recognize the complete failure of life to award contentment. This idea also ties into Schopenhauer's ideas, in *On Religion* and especially in *The World as Will and Representation* (volume II), regarding those practices. It's also a little unusual for Huxley in that he isn't kicking organized religion in the shins with metal-tipped boots, but rather acknowledging a certain sense in these practices. That's probably because he's talking about individual contemplation and choices about engagement with the world, rather than institutions of power and their doctrinal claims.

In earlier writings, Huxley had been gentler or more optimistic. He had always granted that rationality and, specifically, speech, illustrate an existing status for humans that transcends our primate origins. He had implied that people receive a vague karmic compensation during the course of their lives, or claimed that the new standards of scientific thinking had already begun to usher in an age of human happiness. Here, forget it: no such tangible goodness is evident. Our experience is an animal one, because it is experienced by animals, and it's arguably a particularly bad one. Huxley's portrait is doubly chilling given his simple, undemonstrative description—somehow, as a reader, I get that he really, really means it.

In a word, *Evolution and Ethics* was a huge stink bomb tossed into the parlor of human self-congratulation. All the established interpretations of evolution and humanity had been comfortably distributed across the political landscape, each appropriating the terms to consider itself the most evolved. It was still all about privilege and claim to authority, with "scientific" as the new "divine" justification for power. Huxley was having none of it.

However, at the end, his conclusion lapses back into his ideals. In some contradiction to his claim that older cultures were no less mentally astute than modern ones, he posits a social-scientific solution, as yet undreamed of, in the future:

Let us understand, once and for all, that the ethical progress of society depends, not on imitating the cosmic process, still less in running away from it, but in combating it. It may seem an audacious proposal thus to pit the microcosm against the macrocosm and to set man to subdue nature to his higher ends; but I venture to think that the great intellectual difference between the ancient times with which we have been occupied and our day, lies in the solid foundation we have acquired for the hope that such an enterprise may meet with a certain measure of success.

... the highly organized and developed sciences and arts of the present day have endowed man with a command over the course of non-human nature greater than that once attributed to the magicians. The most impressive, I might say startling, of those changes have been brought about in the course of the past two centuries; while a right comprehension of the process of life and of the means of influencing its manifestations is only just dawning upon us ... [*in reference to astronomy, physics, and chemistry*]. Physiology, Psychology, Ethics, Political Science, must submit to the same ordeal. Yet it seems to me irrational to doubt that, at no distant period, they will work as great a revolution in the sphere of practice.

... it would be folly to imagine that a few centuries will suffice to subdue its masterfulness to purely ethical ends. Ethical nature may count upon having to reckon with a tenacious and powerful enemy as long as the world lasts. But on the other hand, I see no limit to the extent to which intelligence and will, guided by sound principles of investigation, and organized in common effort, may modify the conditions of existence, for a period longer than that now covered by history. And much may be done to change the nature of man himself. The intelligence which has converted the brother of the wolf into the faithful guardian of the flock ought to be able to do something towards curbing the instincts of savagery in civilized man. (*Evolution and Ethics*, pp. 83–85[1895 edn.]; pp. 141–143 [1989 edn.])

His language is tentative: he *hopes* for a *certain measure* of success, with hints of social practices or even a re-engineering, which *ought to* result in a behavioral and ethical human profile with more generally happy results. For something that is "irrational to doubt," it's distinctly iffy. My reading is that these final passages do not stand up as argument, but as hope—perhaps a little desperate, at that.

Even Huxley—the man who in *Evidence as to Man's Place in Nature* had publicly demonstrated that humans are apes—couldn't escape the dichotomy between "humans are no beasts," which caricatures humans into quasi-angels, versus "humans are mere beasts," which caricatures nonhumans into selfish savagery. Even casting our real selves into the latter doesn't do it, because the former is still out there to be discovered. His proposal tries to escape the zero-sum between these imagined "beings" by introducing time: *currently*, we are effectively not civilized because we are beasts; *later*, we will not be beasts because we will be civilized. But since the two images remain, Man and Beast as ever, his construct yet retains an indigestible combination of a lowly beast not advancing with a glowing if vague notion of what advancing to true humanity would be (Figure 1.6). The new and actual human experience, waiting in the future, gift-wrapped in all his uncertain words, is still special and exalted as ever.

Human exceptionalism had become a trap, and it held shut into the twentieth century, despite the evidence already described, and in the continued face of

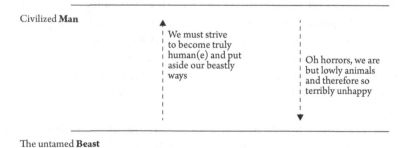

Civilized **Man**

We must strive to become truly human(e) and put aside our beastly ways

Oh horrors, we are but lowly animals and therefore so terribly unhappy

The untamed **Beast**

Figure 1.6 Evolution and Ethics exceptionalism.

all evidence to come. Huxley's peers generally dismissed *Evolution and Ethics* or regarded it as a personal lapse, leaving the distorted versions of evolution intact. In literature, the conundrum underlies the work of the next two generations of Huxleys, but I want to focus on someone else, who did not fear to take the lecture's strongest argument all the way.

Human Animals

Just as Lawrence's lectures were dramatized by Shelly at the conception of modern biology in the early 1800s, so Huxley's *Evolution and Ethics* was dramatized by Herbert George Wells. Whether he was in fact present at Huxley's lecture isn't known, but he clearly knew it or the later publication backward and forward. Wells was perfectly placed to comment on these issues. He was in his late twenties, barely finished with his studies in zoology, during which time he'd been a student of Huxley's, and had been a teacher but was not a practicing scientist. He was also a social activist, although a contentious one and not easily associated with a specific group. His career as an author was recently launched with some essays, short stories, and two books, including *The Time Machine*. He and a few other authors were discovering that speculative fiction could provide a unique and exciting critique of science, values, and society.[12]

This remarkable confluence of all the current questions in one thoughtful, creative young man found its voice in *The Island of Doctor Moreau*, published in 1896. It's inspired by and deeply rooted in Huxley's *Evolution and Ethics*, but—just as with the 1818 *Frankenstein*'s roots in Lawrence's lectures—not to rail against it, but as an informed reflection on the most difficult points that had been raised. Just as *Frankenstein* is not about a feeble-minded, zombified brute, *The Island of Doctor Moreau* is not about the lurking or erupting threat of animalistic savages. Moreau succeeds in his project; his creations are human persons who experience their lives just as we do. The story is instead about what we are like, and the myriad means by which we refuse to see it.

Historically, it captures a moment in the intellectual crisis between biology and the accustomed modes of thought and status. Moreover, it presents the questions as they still need to be asked, free of institutional wrangling and identity politics.

- What is our basis or founding principle for how to treat nonhuman animals?
- Where do our ethics come from, anyway?
- And underlying each of these: What does our status as a zoological animal mean regarding human society, and does it matter?

True to its source, it is also one of the least sentimental novels available, in which its action and butcher-shop horror are not exciting so much as disturbing, wince-inducing, and outright grievous. It is about animal suffering, explicitly including our own. It adds up to the question that drove *Evolution and Ethics* despite its hint of hope: Who are we?

The story is similar to *Frankenstein* in some ways. The scientist, Doctor Moreau, succeeds in creating humans from animals, but interprets the results as failure, and the narrating character, Prendick, recoils from his own understanding of their humanity. In line with the adaptations, Moreau has a subordinated assistant, too, named Montgomery. Its structure is more organized, however (Figure 1.7). Moreau and Montgomery have little or no character arcs at all, so their roles are limited to portraits and then to what their prior decisions and attitudes have brought them. Moreau is physically and intellectually a massive force, certain, unswerving, self-assured, and despite one significant frustration, at peace with himself and his project, whereas Montgomery is depressed, alcoholic, and belligerent, yet connects better with the created people partly out of sheer loneliness. Prendick is the dynamic character, whose knowledge of the situations increases in ways not accessible to the others, and his attitude toward those situation changes several times. He's not a narrator "eye" at all, but takes decisive action more than once. Overtly, it's much more his story than theirs.

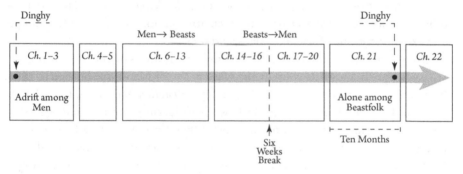

Figure 1.7 Diagram of the plot, *The Island of Dr. Moreau*.

The outer frame is about Prendick being frightened of humans, specifically of seeing what one calls the animal in them. In Chapters 1–3, that means several grim situations, beginning with the temptation and threat of cannibalism, then bullying and cruelty, including being abandoned to die on a raft twice. In Chapter 22, when Prendick is back in England, his fear is much more profound: he can't stand being around ordinary humans doing ordinary things, seeing the animal in every face, every action, and every intellectual, social, and religious endeavor.

The inner frame is about his time on the island, interacting with Moreau's creations, called the Beast Folk. He goes through several phases, first thinking they're humans bestialized by surgery and fearing to share their fate, then, after learning their origins, struggling with different degrees of empathy, and finally descending into terror and hatred. This final full recoil is subtle, because Prendick is a little bonkers at this point. Briefly, despite his fervid imaginings, the Beast

Folk never "revert to the beast" in the sense he imagines, that is, there is no final grunting rampage of violence. Instead, first he can't stand being around the Beast Folk doing ordinary human things, and is then further traumatized when they rather quietly and sadly revert to their original forms. There's plenty of violence in the story—both of the other human characters, Moreau and his assistant Montgomery, meet their ends at the hands of Beast Folk—but it's never actually the berserk "reverted beast released" that's responsible.

Given the complex series of thoughts and emotions in the inner frame, the effect of the final chapter is a remarkable blow—right at the height of Prendick's hatred and terror of the Beast Folk for being too much like people, the story brings him off the island and into his stunned horror at ordinary people, for being too much like animals. In particular, Huxley's purported golden era, somewhere beyond the horizon, in which better-natured humans suffer less in a better society, is nowhere to be found.

In *Frankenstein*, the created man gets his say at two points, with plenty of time to make his points to a committed listener. In *The Island of Doctor Moreau*, the Beast People are no less present and active with their own thoughts and decisions, but they are voiceless in comparison. The four most important barely get the chance to talk at all. Still, enough information is given to work with, more subtly, in order to see what the story is like for them. Despite Prendick being the narrator, I suggest that these characters comprise a shadow-novel of their own.

The structure and the tacit stories are pretty easy to see once one puts aside preconceptions and reads the words with a certain suspicion toward Prendick's narration. Similarly to Victor Frankenstein, his delivery doesn't always accord with what he describes. He develops a considerable fear of the Beast Folks' innate or imminent savagery, for example, but the actual violence and outright malevolence they exhibit is less than that evidenced by the human characters, including himself. The imagery is similarly blurred by what he only hints at or leaves out, and especially by subsequent visual media, which fill in the gaps with material that simply is not part of the book's content.

I don't expect anyone to accept my interpretation at this moment of reading. The novel has been culturally and academically smeared as gaudy and superficial, and it's been thematically reversed, even worse than *Frankenstein* was. Everyone "knows" the Beast Folk are hairy animal-headed monsters, that Moreau is a barking mad sadist, and that it all ends in a rampaging conflagration in which they punish him; that the book is only and ever about how out-of-control scientists do crazy and hurtful things, torturing animals and making monsters who are simultaneously victims and mortal threats to the rest of us. I happen to think it's not, but there's no reason that you should trust me about it. I only ask for a new, close, and thoughtful reading, which I've tried to support by writing this book.

I've done that because the intellectual and ethical crisis of human exception-alism was not resolved in the 1890s. The two social-appropriation distortions of evolution remained. The range of Huxley's and Darwin's actual work remains underappreciated, as opposed to their public lives or the single concept of natural selection. *Evolution and Ethics* effectively failed, leaving behind only its finishing glimpse at utopia, when it might have gone even further in examining the human animal. The political crisis of animal care in research was socially set in place with no grounds for resolution (Figure 1.8).

Rather horribly, this exact lack of resolution became the "new normal" and has only been grown layered with new details and more contorted for over a cen-tury. Biology itself remains schizophrenic about humanity, not yet daring to fold psychology into animal behavior or anthropology into mammalogy. Every other discipline firewalls biology away from it, sometimes with great fear. Ethics and policy regarding nonhuman animals, despite some gains, remain barely coher-ent. Education, legislation, and practices concerning our biological identity have twisted and turned, but the underlying exceptionalism remains fixed. *The Island of Doctor Moreau* was not only the most informed and challenging book of its time, but it is still the single most relevant and provocative book about these issues for *our* time, because it throws the still-unresolved crux of human identity and excep-tionalism for these precise controversies into plain sight.

Readings

My primary reference is obviously Marilyn Butler, *Frankenstein and Radical Science* (1993), which is the title for her introduction to current reprints of the 1818 ver-sion of the novel, and also her *Romantics, Rebels and Reactionaries: English Literature and Its Background, 1760–1830* (1998). *Making Humans* (2003), edited by Judith Wilt, includes the 1818 version and Huxley's *Evolution and Ethics*, but does not include Butler's analysis and does not mention Lawrence.

Mary Shelley was a literary genius, and fortunately many reviews are available. Useful biographies include Anne K. Mellor, *Mary Shelley: Her Life, Her Fiction, and Her Monsters* (1989); Joan Kane Nichols, *Mary Shelley: Frankenstein's Creator* (1998); Miranda Seymour, *Mary Shelley* (2011); Dorothy Hoobler and Theodore Hoobler, *The Monsters: Mary Shelley and the Curse of Frankenstein* (2007); Muriel Spark, *Mary Shelley: A Biography* (1987); and Roseanne Montillo, *The Lady and Her Monsters* (2013).

I've used the Princeton University Press version of *T. H. Huxley's Evolution and Ethics*, published in 1989, including James Peralis's essay on its historical context and George C. Williams's discussion of human behavior and evolution-ary theory. Useful biographies include Adrian Desmond, *Huxley: From Devil's Disciple to Evolution's High Priest* (1994); Paul White, *Thomas Huxley: Making the*

—	1802	William Paley, *Evidence of the Existence and Attributes of the Deity*
—	1809	Jean-Baptiste Lamarck's *Histoire Naturelle*
—	**1816–1818**	**Lawrence's Hunter lectures**
—	**1818**	**Shelley, *Frankenstein*, reprint 1822**
—	1828	George Combes, *The Constitution of Man*
—	1844	Robert Chambers (anonymous), *Vestiges of the Natural History of Creation*, first edition
—	1847	8th and final edition of *The Constitution of Man*
—	1851	Spencer, *Social Statics*
—	1859	Darwin, *On the Origin of Species*
—	1859	*Vestiges* discussion of the *Origin*
—	1860	Huxley's review of the *Origin*
—	1860	Oxford evolution confrontation between Huxely and Wilberforce
—	1861	First actual gorilla displayed in Europe (stuffed)
—	1862	Spencer, *First Principles*, "survival of the fittest"
—	1863	Charles Lyell, *The Geological Evidences of the Antiquity of Man*
—	1863	Huxley, *Man's Place in Nature*
—	1864	Huxley, Spencer, and others form the X Club
—	1868	Ernst Haeckel's *Natürliche Schöpfungsgeschichte*
—	1869	5th edition of the *Origin*, including phrase "survival of the fittest"
—	1870	Alfred Russel Wallace, *Contributions to the Theory of Natural Selection*
—	1871	Darwin, *The Descent of Man*
—	1872	6th and final edition of the *Origin*
—	1872	Darwin, *The Expression of Emotion in Man and Animals*
—	1876	English translation of Haeckel's book as *The History of Creation*
—	1879	Spencer, *Data of Ethics*
—	1884	12th and final edition of the *Vestiges*
—	1887	W. S. Lilly, *Materialism and Morality*
—	1888	Huxley, *Science and Morals*
—	1889	Wallace, *Darwinism*
—	1891	First fossil findings of *Homo erectus*
—	**1893**	**Huxley, *Evolution and Ethics*, published 1895**
—	**1896**	**Wells, *The Island of Doctor Moreau***
—	1904	Wallace, *Man's Place in the Universe*

Figure 1.8 Timeline of human exceptionalism.

"Man of Science" (2002); and Sherrie L. Lyons, *Thomas Huxley: The Evolution of a Scientist* (1999). An account of his views of humanity is included in Misia Landau's *Narratives in Human Evolution* (1991) in comparison with Darwin's and Haeckel's.

Bruce Moran's *Distilling Knowledge* (2006) provides an accessible history of chemistry as context for reading William Lawrence, *Lectures on the Physiology, Zoology, and the Natural History of Man*, delivered at the Royal College of Surgeons (3rd edition, 1823).

I've used the 1996 Dover Thrift edition of *The Island of Doctor Moreau*, which is an unabridged reprint of the original 1896 publication by William Heinemann, London.

There are too many biographies and reviews of Wells to cite completely. A good start includes Norman and Jean MacKenzie, *H G Wells* (1973), which is especially helpful regarding his influence by Huxley; Vincent Brome, *H G Wells: A Biography* (1951); and Michael Sherborne, *H. G. Wells: Another Kind of Life* (2010).

Notes

1. The term "materialism" has undergone many twists and turns of meaning. In 1810–1830, relative to Lawrence, I've found it better to use "material" for the adjective, to indicate strict physical causes, rather than the technically correct "materialist," given the implications developed later for that phrasing.
2. William Lawrence, *Lectures on Physiology, Zoology, and Natural History of Man*, 1823, published by James Smith (available at https://archive.org/details/lecturesonphysio00lawrrich).
3. Two other difficult terms are "abiogenesis" and "biogenesis," which at first meant what they sound like, with the former indicating a chemical underpinning to life. However, they flipped or reversed meanings later in the nineteenth century, relative to the debates of whether living things were produced *de novo* from nonliving material or from other living things. It was difficult to address the question, if life is nonvitalist (no super-added forces), then why it only emerged from other living things, and these terms twisted quite a bit in the winds of that discussion.
4. The three formal components of cell theory are that living things are composed of one or more cells, that cells are the smallest unit of life, and that cells arise from prior cells. Neither Antonie van Leeuwenhoek nor Robert Hooke discovered cells in terms of this understanding. They *looked* at them, and named them as things they saw, which is very important but isn't to the point. Cell theory begins intellectually with Lawrence, then booms with Schleiden, Schwann, Pasteur, and others, including Huxley's "primordial fluid." Biology is no less guilty than other disciplines in wanting to backdate its modern content further than it should.
5. Much of Lawrence's text about humans is ethnically tagged and includes racist concepts and terms. Appalling as this is, I think it does not interfere with many of his points when they are applied, as they do, to every human being.
6. Friedrich Nietzsche, *The Gay Science*, second edition, 1887, translated by Walter Kaufmann, published by Vintage Books, 1974.
7. The chronology of *Frankenstein* publication is tortuous: the first version was published anonymously in two or more short volumes (a common presentation at the time) on January 1, 1818; then the stage play *Presumption: the Fate of Frankenstein* by Richard Brinsley Peake was presented within a year or two. In the play, the creation is mute

and infantile; Frankenstein has a comedic assistant named Fritz; and Frankenstein collapses in shame, remorseful of his "impious labor." I suspect the play and similar productions were a constant feature throughout the nineteenth century, contributing to later film tropes. The novel was then republished in 1822 under Shelly's name with emendations by her father, again in short volumes. Its 1831 publication puts the whole story between two covers for the first time, including the revisions and additions as described by Butler.

Both Lawrence and Shelley succumbed to social pressure, but one should remember the terrible fate awaiting Victorian families, especially women, once targeted for ostracism.

8. Mary Shelley, *Frankenstein: or, The Modern Prometheus*, The 1818 Text, published by Oxford World's Classics, Oxford University Press, 1994.

9. Robert Chambers, *Vestiges of the Natural History of Creation*, 1844, published by John Churchill (available at https://archive.org/details/vestigesofnatura00unse).

10. Thomas Huxley, *Evolution and Ethics*, 1895, published by Princeton University Press, 1989.

11. In *The Devil's Dictionary* (1911), Ambrose Bierce defines accident as "an inevitable occurrence due to the action of immutable natural laws," which seems admirably suited to Huxley's explanation of the history of life and specifically the origins and current status of humans.

12. Harlan Ellison suggested the term "speculative fiction" to replace "science fiction" and its abbreviations, like "sci-fi," and it fits well here.

2

The -ism that Wasn't

Moreau, Montgomery, and Edward Prendick are the three primary human characters in the story. They're all English, although if the names are any indication, they have slightly different local ethnic histories: Norman, Scottish, and Anglo-Saxon, respectively. They are minimally described in terms of appearance and other details—no first names are provided for the first two, for instance. Moreau gets the most description: he's a big man in his sixties, in excellent, even startling physical condition, with a full head of grey hair. His history is explicit: a noted surgeon and outspoken researcher, driven from England a decade earlier when accused of cruelty to animals, with evidence. Apparently he never married and had no family life worth mentioning. He uses no patronizing upper-class British mannerisms or phrases. He is calm and certain in his current life and in his engagement with his project.

Montgomery is about thirty, with flaxen (blond) hair and either a habitually stupid expression or some skill at putting it on. Eleven years previously he had been a medical student at the University College in London, "very average" as Prendick judges, and he speaks of his studies dismissively, preferring to remember the nightlife. He refers not too obliquely to a "mistake" apparently bad enough to ruin his life, almost certainly part of said nightlife and worth arrest and conviction. He was effectively rescued by Moreau as the latter fled England, and considers himself unable to return. He has a bad conscience about these events and about the current surgeries on the animals; he is lonely and full of self-pity. He drinks heavily and has a rotten temper.

Prendick is in his mid-twenties, shorter than Montgomery, physically resilient and evidently skilled at running and at shooting, perhaps from his background of "comfortable independence." He may be the highest in the British social hierarchy among the three. He is also an abstainer "from birth," which is to say, brought up in temperance activism and therefore in committed progressive circles. This background is consistent with his reading an anti-vivisectionist pamphlet titled "The Moreau Horrors" as a boy, when the researcher came under criticism. He was also a student at the University College, but studied zoology with Thomas Huxley rather than medicine.

One thing should jump out: *they're all scientists*, trained at or at least associated with the same institution. They know the same neighborhood and general educational culture. This is a pretty tight knot of people from a specific time and place. Their differences are found in their relationships with science itself, its practices, attitudes, and internal conflicts, from within.

And you can't get more embedded in the internal disputes of science than the intersection of interventional animal use, human subjects (effectively), evolutionary theory, the definition of humanity through function, and the dynamics of human social interaction. Our own science can't even permit these things to be unified professionally and intellectually, as witness evolutionary biology, mammalogy, anthropology, sociology, and psychology, among others. The three characters in the story, being who they are, explicitly strike at some of biology's most sensitive fracture points. Specifically,

- Moreau is the research physiologist, to whom pain and any other consideration are secondary to his questions—but those questions are so rooted in the Man/Beast divide that he'll never see a real person he creates as truly human.
- Montgomery is the medical student, profligate, immature, a bit pathetic, and wondering how in the world he ever did what he did, or came to this.
- Prendick is the zoological idealist, appreciative of science as knowledge, hopeful for the future of humanity, and completely unprepared to deal with real humans when they aren't what he wants to see.

Huxley's *Evolution and Ethics* doesn't mention the research physiology world, a strange lapse considering how much authority he had held in that sphere and how much educational reform he'd spearheaded. Surely the treatment of animals in that discipline would have been a perfect topic for his musings about "ape and tiger" behaviors and morality.

The novel has no such lapse. Here, the interplay among culture, education, scientific questions, scientific practice—and the core role of humanity and animal as separate things—is laid bare. Dissected, one might say.

It Does Not Please You

I keep talking about the Man/Beast divide as a nineteenth-century topic, but oddly, the dichotomy had already been irrevocably smashed over a century before, even made scientifically obsolete, in an obscure branch of natural history.

Opening the Door

It goes back as far as 1735. The Swedish scientist Carl Linnaeus's *Systema Naturae* needs little introduction or description; it's a primary text of biology and may hold the longest record in that discipline in terms of pure procedural technique, as his methods are used to this day. At first glance, it looks like a straightforward list of living things, certainly impressive, but sort of boring and a bit obsessive-compulsive—until you see that the list isn't arranged into a stepwise-ladder, but a sprawling array of nested ones. There isn't any Great Chain of Being there.

I like to explain it as boxes inside boxes. Open the chest of living things, and you see a bunch of big boxes. Open one, and you see several more boxes in it, and a given box may have many or few boxes in it, or even just one. A given creature is not only itself, it's the boxes it's contained in. You're a human, but a human is a kind of primate, and all primates as a group are a kind of mammal, and all mammals as a group are a kind of vertebrate, themselves as a group being a kind of animal. The generic name for a "box" of any size or contents is *taxon*, and giving every box a name is called *taxonomy*.

Taxonomy is the most consistently dynamic discipline in biology. New creatures are always being found, for one thing, but more important, any group (box) is always subject to revision, the technical word for figuring out whether the current ordering of the boxes is really making sense. Taxonomists distinguish between alpha naming, which is simply throwing a name and provisional box membership onto a newfound species to get it into the system of boxes in the first place, and revision. Revisions are grounds for disputes about proposed renamings, to put it mildly. Despite being less glamorous than geneticists or climate researchers, taxonomists have a well-deserved reputation as the fussiest and fiercest arguers in biology.

No one can say it's not justified, though. Evolutionary theory, as a discipline, is nothing but debating the organization of the boxes and more debating about how they got that way. Linnaeus wasn't writing in the context of evolution; that concept would not be articulated and named until about 1800, after his death in 1778. But my goodness, did it apply. Even as the term "evolution" was coined, Jean-Baptiste Lamarck tied it to the concept of lineage, conceiving of the history of living things as multiple transformations, a series of branches. In this, he deserves recognition as one of the discipline's foremost intellectual contributors: the diversity of life at any given moment is the outcome of all the prior branchings, from most ancient to the most recent. The modern word for this is *phylogeny*, informally "evolutionary tree," although it's really a trunk-less bush, and even as Lamarck put down his pen, it was wedded to the existing Linnaean practice of taxonomy. Today, this marriage is called *systematics*. Without it, the rest of biology as we know it is no science at all, which Lawrence recognized and made the centerpiece of his lectures, urging that the chemistry of life be compared among any and all living creatures as organized by taxonomy.[1]

Linnaeus's emphasis on shared distinctive features remains the central technique of systematics. Today, the most influential reference for making phylogenies is Willi Hennig's *Phylogenetic Systematics* (1966), which refines but does not alter this principle.

Here's how it works. When Linnaeus named a group, he wasn't talking about some distinct feature that a single species may have. Every species has one or more of those, by definition. He's talking about a distinct feature that is *shared* by two or more species, which is used to name the taxon (box) containing them. The larger, enclosing groups are designated using the same logic.

Now for the earthquake shock: he included humans in the arrangement. Not in our own non-animal category, either, but right in there among the primates, which he called Anthropomorpha in the first edition.

It's clear from correspondence that he knew exactly what he was doing. His exchange with Johann Georg Gmelin is directly available through national archives in Sweden and Russia. From his letter to Gmelin in 1747:

It does not please [*object unspecified, arguably "you"*] that I've placed Man among the Anthropomorpha, perhaps because of the term 'with human form,' but man learns to know himself. Let's not quibble over words. It will be the same to me whatever name we apply. But I seek from you and from the whole world a generic difference between man and simian which is in accord with the principles of Natural History. I absolutely know of none. If only someone might tell me a single one! If I would have called man a simian or vice versa, I would have brought together all the theologians against me. Perhaps I ought to have by virtue of the law of the art [*discipline*]. [*Minor variations in translation from his Latin exist; none changes the meaning.*] (Linnaeus, letter to Gmelin)[2]

This is what's called a smoking gun. Considering how elusive they are in the complicated dialogues of science, it might be one of the most significant. Let's break that text down into parts, if you'll forgive my audacity in paraphrasing the father of taxonomy.

It does not please [*you*] that I've placed Man among the Anthropomorpha, but man learns to know himself.

My paraphrase: Get over it, Johann. We're not talking about something that's subject to your or my feelings. The whole point was to examine ourselves, and if we're doing this honestly, then we have to take it as we find it, not as what we might find most pleasing.

Let's not quibble over words. It will be the same to me whatever name we apply.

My paraphrase: Don't pretend this is about the exact term "Anthropomorpha." If it were, I'd change the name to something you like. You're upset about what the group really includes, not what we call it.

I seek from you and from the whole world a generic difference between man and simian that [*follows*] from the principles of Natural History. I absolutely know of none. If only someone might tell me a single one!

My paraphrase: If you want to have some kind of different grouping, then you're going to have to show me one, single, actual feature among the nonhuman simians that they share and humans do not. Go ahead. Try.

If I would have called man a simian or vice versa, I would have brought together all the theologians against me. Perhaps I ought to have by virtue of the law of the discipline.

My paraphrase: And you know what? I actually already did soften the blow
by giving humans any distinction from the nonhuman simians at all. If I'd stuck
by my techniques, then they'd all go into one group together. I didn't because
I feared what people like you would say, and I caved. I shouldn't have.

That makes two gunshots, actually. First, he deliberately distorted his conclusion in fear of repercussions, a distortion that has so far resisted clarification outside very specialized science. Second, his methods—if applied as he'd used them
regarding every other known living thing—conclude that humans are—wait for
it—apes (not "from," not "near," not "closest cousins," but *are*).

Linnaeus was right on the button. Then and now, you can list distinctive features of any single species in the large tailless primates all day long, which doesn't
tell us anything. You can also list a raft of distinctive features shared uniquely
among them, including ourselves, and that is important, because that makes them
a group, or in evolutionary terms, relatives. What you can't do is find a feature that
is shared by the tailless primates we don't call human, but that humans *don't* have.
You can't make a group of "them" without us in it (Figure 2.1).

Linnaeus received criticism from his peers in the sciences, but he was never
hit by the kind of outraged horror later directed at Lawrence and many others
by an entrenched power structure. Perhaps his deliberate obscuring of his most
teeth-rattling point was successful toward that end, or perhaps he was lucky in
that the larger context of evolution had not yet been articulated or politicized.
Systemae Naturae was a catalogue, not a process argument. However, by the mid-
1800s, his bravery to include humans at all, and right there in the Primates, had

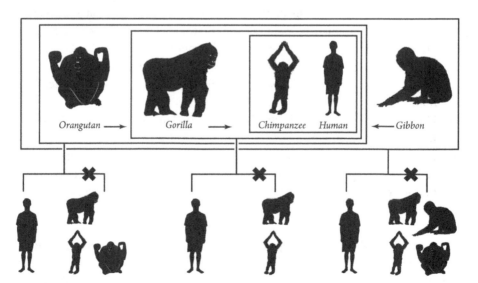

Figure 2.1 Primate classification.

done its work: the door leading out of the Man/Beast divide was already open and unavoidable. Huxley and Darwin are often incorrectly credited with that achievement, but although that's not true, they swung it wide and walked through, in a series of stunning works:

- Huxley's "Lectures to the Working Man" series, leading to his *Evidence as to Man's Place in Nature* (1863), presented his correct observations regarding apes, consistent with Linnaeus's admission in his letter: first, that humans and gorillas, and tacitly, chimps as well, share far more in common than either does with monkeys, in features of the hands, feet, and brain, among others; second, not his main point, but still explicit and critical to my argument, that it is not logically possible to group gorillas and chimps as a distinct pair away from humans, based on anatomy. At this time, Huxley carefully distinguished human abilities as distinctly superior to those of other animals, despite a humble origin.
- Ernst Haeckel's *Natürliche Schöpfungsgeschichte* (1868), or *The History of Creation* (English, 1876), introduced embryological comparisons and crucial terms, including phylogeny.
- Darwin's *The Descent of Man* (1871) presented his ideas about humans as products of evolution, very distinct from the "apex of creation" model so familiar from the *Vestiges* and in Spencer's writings; it also introduced the idea of sexual selection and specifically that humans have been subject to it. Its content regarding morals and sociality as animal features, and thought as a "secretion of the brain," is no less than extraordinary.
- Darwin's *The Expression of the Emotions in Man and Animals* (1872) demonstrated that human emotions and expressions are homologous with those of other mammals, in retrospect one of the most ambitious published challenges to human exceptionalism, even relative to Huxley's work. These insights were off-message from what everyone "knew" evolution and so-called Darwinism were, receiving no critical attention and, unfairly, being less read even today. Much of Darwin's research also concerned the decision-making capacities of nonhuman animals.
- Huxley's "On the Struggle for Existence in Human Society" (1888) and *Essays upon Some Controverted Questions* (1892) led to his *Evolution and Ethics* (1893/1895), as discussed in Chapter 1: breaking with the idea that humans are currently excellent and perfected by evolution. In this, he downgraded his prior positive view of humans' current social and intellectual elevation drastically.

As with all scholarship from a past era, there's a lot to wince about and a lot to acknowledge as altered in today's body of theory, but their essential points regarding the biological status of humans still stand, solid and quite fearless. Effectively

Linnaeus's unpublished specification and Lawrence's unvarnished but buried points came out into the light: humans are apes—anatomically, phylogenetically, and behaviorally.

Slamming It Shut

Given these points by Linnaeus, Darwin, and Huxley, who stand among the most influential biological thinkers in history, one might naively infer that these insights were adopted by the new and rapidly unifying discipline, resulting in an easy "science thinks this, everyone else thinks that" version of the dichotomy. Everything about biology was coming together: cell theory, physiology, biogeography, evolution, taxonomy, behavior, embryology, and paleontology, with genetics only moments away—and everything about humans was snapping into place along with every other creature, detail by detail. Biology seemed about to coalesce into a remarkable culture of shared non-exceptionalist understanding.

However, that did not happen. People had too many agendas and uses for Darwin-*ism* as a *directive* for them to apply natural selection or Darwin's other work as actually written, and for them to process these explosive insights even a little. The later professional context for biology and education followed this path as well, to slam that door shut again and extract profit through the mail slot, to be discussed further in Chapter 10.

It hadn't helped that Linnaeus never did rouse his indignation enough to make it stick in print. In the tenth and final edition of the *Systema Naturae* (1758), he placed humans even further distanced from the other apes. They remained in the Primates, and listed with the other apes, but not among them. This taxonomic point isn't a mere detail; it's the crux of a remarkable split concerning evolution that has persisted ever since. During the 1800s, it's evident between, on the one hand, a very few scholars like Darwin and Huxley, and, on the other, almost all of the other leading intellectuals. They weren't and aren't going to consider humans to be animals, using "from" and "not only" as a way to escape.

Most of the gaudy wrangling about evolution, God, humans, fossils, creation or Creation, and who knows what else is a distraction or useful political code, and always has been. I maintain that no one is really upset about evolution and natural selection as such, but instead about humans being animals. The upshot of the nineteenth-century processing about evolution and Darwinism was to shove that question under the rug—ignoring Linnaeus, ignoring much of Darwin's and Huxley's actual scientific work, subverting the term "Darwinism" into various fibs about humans being awesome in one way or another, or assigning it demonic social content as a target.

Huxley's *Evolution and Ethics* deserves a second look in this context. It has nothing to do with his long history of butting heads with authoritarian churchmen; it's a serious reflection on what humans are like and how they feel, if they are in fact animals, and what that has to do with a less unjust society. He made it almost all the way, reserving his hope for a better human world as some unknown activity in the future.

I can list the reactions to the concept "humans are animals" quite simply; they may seem different, but are all really versions of recoiling into human exceptionalism.

- We never were animals, ever ever.
- We may have been, but we aren't now.
  - Except for some of us, who need to be elevated or kept in their place.
- All right, we're animals, but only technically, it doesn't mean we're *like* animals.
  - So don't be so beastly, like you constantly keep doing.
- Or conversely, animals are just like us and therefore are as elevated and noble as we are.
  - So we mustn't treat them "like animals."

It's hard to abandon the Man/Beast dichotomy and say that we are animals, then, now, and forever—without the caricature of "animal," the dull or frenzied Beast, creeping into it, either as a looming threat or as a tempting indulgence. You'd have to say that we're animals, then, now, and forever, and also admit that it carries no directive or instructive force—that we have no idea what it *means*.

Politically, I think this was Huxley's last, effectively failed struggle to recover natural selection from its co-option into *either* imperialism and racism *or* from the self-congratulatory assumption that it had already produced the virtuous human. This lecture is probably his most explicit attempt—I can't help but think of it as a desperate attempt—to establish a working common ground, without fantasies, between the evolutionary naturalists and the culture of social reform.

Darwin in the Middle

In the decade of *The Island of Doctor Moreau*, biology as it is organized today had almost completely come together. All through the nineteenth century came a series of intellectual shock waves built of new inquiries and continuous debates, but its final decade also brought its *social* tipping point, establishing a new place for biological science in education, professions, funding of research, policy, and technology.[3]
How did any and all of this relate to the rest of society? Consider these:

- Imperium: the end of the Napoleonic wars and the rise of the Pax Britannica, true Victoriana, and the last and broadest phase of classic colonialism, politically and technologically grading into the wars of the twentieth century

- Revolution: specifically the uprisings of 1848, which framed the global politics of labor for the next sixty or seventy years
- Nationalism: the appearance of a new definition of "state" and citizenship
- Industrial Revolution: first the long leading edge of agriculture and transport, then the appearance and rapid development of powered systems.

None of these would seem directly affected by intellectual details about cells and skeletons, but there's more connection than you might think. Biological science as we know it, professionally and intellectually, arose in this ferment, but also contributed to it, wrapped up with every crisis issue.

- Technological: war, transport, agriculture, energy, communication, and medicine
- Public health: policy and reform
  - The context of truly horrific disease outbreaks, casting physiology as a quickly applied body of knowledge, creating an associated industry of experimental medicine
- Government funding: organization and support for technology development, exploration, research
- Social organizing: the middle or merchant class developing its position of power
  - Abolition (slavery ended in Britain in 1833), temperance, political representation including women's suffrage, inheritance laws, sexual politics, and much more
- Religion: the reconfiguring of church power relative to state power
  - Profusion of new, politically active Protestant sects
- Education: reform and national policy, specifically the appearance and reinforcement of new professions.

Among the hottest words of the time were "evolution" and "natural selection." Rather than flat scientific bits of natural history, they became the touchpoint for a broad, uneasy shift from the metaphysical to the physical perspective. It was apparent in all kinds of conceptual conflicts: nature and the natural, religion and ethics, the role of technology in power, and the human treatment of nonhuman animals. This shift was above all *incomplete*, in each case breaking upon the rock of human-animal exceptionalism, and then receding. Darwin's ideas, and especially the broader potential of evolutionary theory, became casualties.

Evolution First, Then Selection

Few, if any, of my students, biology majors included, have come into class knowing that the concept of evolution existed before Charles Darwin was born.

Some of the relevant names and works include le Comte de Buffon's (Georges de Clerc) *Histoire Naturelle*, published 1749–1788; Erasmus Darwin's *Zoonomia* from the mid-1790s and *The Temple of Nature* (1803); and Jean-Baptiste Lamarck's *Philosophie Zoologique* (1809), among other works.

The term's exact author isn't clear, which reflects the concept's broad, diffuse presence in the intellectual circles of natural history, and later, in the larger society through the *Vestiges*, the remarkable publication that I mentioned in Chapter 1. There wasn't so much a single person and a single textual idea, as an array of public and private position pieces and debates. In this discussion, evolution was conceived as a ladder of improvement, in which humans held pride of place as its achievement, even its purpose. The long-standing name for this linear hierarchy is the Great Chain of Being, but Lamarck's presentation and the *Vestiges*, despite their differing political associations, both tweaked it with the concept of change from one form into another, through processes worthy of investigation, and including humans in those processes.

Today, among specialists, the term "evolution" is much more boring: nothing but organisms' history of change, technically called *descent with modification*, with its core concept being homology. It's carefully tagged as an outcome with quite a few different causes, including natural selection, but no single one defining it. There is no progression or trajectory; and the familiar image of the tree of life is better conceived as a trunkless bush. It doesn't engage at all with heady abstractions such as "where it's all going" and "how it all began" and "what's it all for," to the extent that when I say this even today, nearly anyone I'm talking to looks frustrated and says, "No, no, you know what I mean!" and yes, I do—they mean what the *Vestiges* was talking about, and culturally, that seems to have been fixed in stone.

The most famous of the several causes within evolution is natural selection, conceived in the 1840s by Charles Darwin and published in 1859 as *On the Origin of Species*, in tandem with work by Alfred Russell Wallace, and as vigorously supported by Thomas Huxley and Ernst Haeckel. It was published during a very hot decade for evolution, in which it became more visible and perhaps more provocative than it otherwise might. The 1853 edition of the *Vestiges* had contained the still-anonymous author's first rebuttals to criticism, so was probably its most exciting version, providing a run-up of popular debate. The really explosive publication arrived in the same year as the *Origin*, called *Essays and Reviews*, edited by John William Potter, featuring seven clergymen's pro-science positions in open challenge to the establishment doctrines of materialism-as-blasphemy and miraculous creation. That was so shocking as to prompt a genuine trial for heresy, which underwent conviction and appeal. In 1860, Huxley gave a public, vocal face to the *Origin*, and to evolution in general, in his review and in publicly defying Bishop Samuel Wilberforce in a debate at Oxford, and the same Wilberforce would go on to seek conviction for the authors featured in

Essays and Reviews. Soon, too, a new edition of the *Vestiges* appeared, including a discussion of Darwin's book.

I go into this detail because the *Origin* became much bigger than its actual topic, through its heightened visibility in this furnace of public discussion. It became a magnet for every imaginable worldview looking for a world to support it, and almost instantly, this magnet had a name. Huxley coined the term "Darwinism" in his review, which in retrospect was a bad idea, implying that there could be an "ism," a purpose, which this purely physical-scientific phenomenon supported. The term was immediately compromised, as everyone wanted to validate and modernize their hobbyhorse by hitching it to this new Darwinism thing, and almost immediately, everyone disagreeing with them wanted to demonize it. (Huxley had bad luck with his coinings; his later term "agnostic" has also been twisted in its adoptions.)

Views

If one phrase could summarize nineteenth-century thought, my choice would be, *explain the world with humans in it, and construct society accordingly.* It was an age of reorganizing power in a furnace of empire, mass action, inordinate wealth, and sea changes in technology, and everyone pouring effort into it wanted a metaphysical and scientific Idea to back up his or her position.

These efforts coalesced into specific combination of ideas, aims, and activities, which I'll call a *view* to emphasize the importance of their starting assumptions (Figure 2.2). I think the relevant views of the mid- and late nineteenth century can be summarized in two sets:

- Spiritual and ideal directives
  - Organized religion
  - Reform activism
- Physical world
  - Science branding, identification as "scientific"
  - Physical inquiry, or technical science
  - The evolutionary view is diverse among, roughly, Lamarck's evolution, Chambers's development, Darwin's natural selection, and Spencer's interpretation

Figure 2.2 shows the views as they relate to one another, but it doesn't describe the people. Differing views are sometimes described as camps, which I think gives the wrong idea that people and groups "fall into" camps and become unswerving adherents, disconnected from one another. That's not how it works at all. Instead of camps, the views are instead like confused bats flapping around, variously seized upon and combined in different ways by different individuals and organizations.

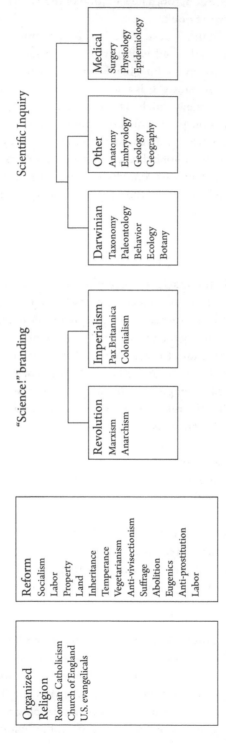

Figure 2.2 Diagram of views.

Think of a person or group with a definite goal cherry-picking across them and reinterpreting their internal phrases to make a working mash-up.

"Darwinism" as a term, and a term only, was solid gold for this purpose. Appropriating this modern, scientific, provocative, media-drawing -ism idea became a priority for any number of these mash-ups, especially in combination with various contemporary meanings of evolution. That's why there isn't a single "Darwinism" but rather a whole bunch of contradictory constructs, each distorted, each tied to specific social values and efforts that combine two or more of the views, and collectively eclipsing the actual text.[4]

The most extreme example is found in the person and writings of Herbert Spencer, a British philosopher of sorts, pundit, social commentator, political theorist—basically a celebrity intellectual, one of those guys who's regarded as doing the thinking for the rest of us. You can't find a better example of the type than Spencer: prolific, authoritative, placed at the intersection of journalism, academia, and policy, perceived as provocative yet somehow never straying from the establishment teat. In his *First Principles of a New System of Philosophy* (1862), he effectively absorbed the term "Darwinism" into the Great Chain of Being framework and, crucially, he subverted the book's technical argument with his phrase "survival of the fittest." Bluntly, this phrase is *not* part of Darwin's ideas, but rather Spencer's own, which he seems never to have realized.

Nearly everyone had synonymized Darwin's ideas with evolution as it was already popularly conceived, as a cosmic improvement program, including the "final" step of humans internally evolving progressively as well, eventually to produce the perfect man in the perfect society. I read the successive editions of *On the Origin of Species* as a tug of war between Darwin and many appropriations, but especially Spencer's, who eventually found his way into it, when the fifth edition featured "survival of the fittest," phrased rather grudgingly by Darwin, as it seems to me. Spencer's subversion and phrasing effectively permanently replaced Darwin's writings, culturally speaking. Today, nearly everything repeated or referenced as "Darwinist" or "natural selection" outside the discipline itself should be called "Spencerism."

"Spencerism," if we may call it that, was most accommodating to appropriation. His survival of the fittest terminology lent itself instantly to the celebration of ruthlessness and privilege, although that was not the most common usage until very late in the nineteenth century, including its association with industrial monopoly, imperialism, and racism. During most of the time period I'm discussing, his name was more often invoked differently. [5] Perhaps more surprising to the modern reader, evolution and Spencer specifically were frequently invoked in activism and reform, often integrated with religious networks and even in secular efforts toward the crucial goal of moral uplift. This "Darwinism" was a benevolent improving and harmonizing process, a particular *moral* way to fold it into policy. In this construction, humanity's good fortune in being evolution's most perfect product also brought its best members the *noblesse oblige* to take care of those who weren't quite "there" yet, whether they were more lowly humans or other creatures.

Significant personalities among these evolution-oriented reformers include:

- Alfred Russell Wallace was an outstanding naturalist and geographer, a dabbler in spiritualism later in life, and an influential activist concerning land reform; he was an avowed socialist and, unlike most of his contemporaries, opposed eugenics. He was Darwin's parallel or partner in arriving at the idea of natural selection, but he eventually differed with Huxley and Darwin in retaining the *Vestiges*-type human-centered, progress-focused view of evolution. Significantly, he is probably responsible for the inclusion of Spencer's survival-of-the-fittest terminology in the fifth edition of *On the Origin of Species*.
- Minot Judson Savage was an American Unitarian minister, activist, and author who exemplifies the positive, progressive casting of Darwinism as interpreted by Spencer. He developed his ideas of "Darwinian evolutionistic optimism" in a considerable number of essays, sermons, and books. Less immediately spiritual than Wallace and other similar thinkers, Savage thought of natural selection as undirected in the moment, but "naturally" tending to diminish meanness or selfishness, and to favor the social and experiential benefit of all. Huxley tended toward these views during the 1860s and 1870s.
- Francis Power Cobbe was an Irish author, activist, and political organizer, one of the key figures in the legal history of animal welfare, women's suffrage, property rights, and related matters. She was strongly influenced by the *Vestiges* discussion, specifically in evolution as a force for collective improvement and increasing justice in society, but eventually disavowed Darwinism specifically, as "survival of the fittest" came to be synonymized with selfishness and ruthlessness.

That's where Prendick comes in. He's been raised in temperance, educated in zoology by Huxley no less, and is generally inclined to think that people have evolved *from* animals *into* something much better, and that such social action as temperance is the next step in that process. As a young man of means without much adversity in his life so far, he has been raised in and exposed to these ideas, but has not seen them tested—or perhaps, has not yet seen what he himself is capable of doing.

Two Windows

Darwin and Huxley were not detached intellectuals at all, but were right in the thick of cutting-edge policies, as in the boiling conflict over animals used in research between reform activists, especially Francis Power Cobbe, with the culture of medical physiologists, which had itself become a scientific and political powerhouse. It was a perfect storm of clashing views, human identity, human relation to nonhumans, science in action, and social policy, in which "Darwinism" as a term, such as it was, as well as Darwin's and Huxley's own social role and ideas, became the object of a tug of war.

According to Patrick White's novel *The Vivisector*, published in 1970, if you were to wander down London's Oxford Street in the early twentieth century, you would encounter two shop windows, side by side. In one, displayed by the Anti-Vivisection Council, was a realistic model of a dog strapped to an operating apparatus, effectively crucified, obviously in agony, its abdomen open and guts protruding. The display was accompanied by a number of pamphlets and placards. The other window, right next to the first, represented the Research Defense Society. It displayed photographs of famous researchers like Louis Pasteur and one of a smiling mother and daughter, including the surgically precise caption, "Which would you save—your child or a guinea pig?"

White's novel is not otherwise about research on animals, and its two windows may be pure symbolism, but this bit of imagery nicely captures history. The organizations are real, although the first was named the National Anti-Vivisection Society (as renamed from the Victoria Street Society in 1897). The windows in the fictional scene represent a real-world knot that had been slowly, inexorably positioned throughout the nineteenth century, tied in 1875–1876 during the Anti-Vivisection Bill debate, and just after the period of the novel, soaked in salt water by the Brown Dog riots of 1907.

From this point, discussions of experiments using nonhuman animals were and are guaranteed to founder. Each window only exists as a moral rock to be thrown at the other, anchored in "facts" that are merely historical artifacts. It's a grossly anti-intellectual, misinformed, and hardened *folie à deux* that has persisted until this day—how did it come to this?

Experimental research using nonhuman animals and activism on their behalf developed in tandem. Chapter 4 presents a more thorough description of the former; here, suffice to say that the practice had developed into a full-fledged industry by mid-century, under no regulations at all, and was especially notorious concerning the University College, located on Gower Street in London (see Figure 2.3 for timeline of related events).

This is Moreau to the letter, built rather exactly from several famous physiologists of the time. He's squarely in the physiologist view, which includes the immediate application of science as progress, almost synonymizing it with technology. He's tied to the Darwinian view as well, just enough to mine it for questions and ideas, but also to dismiss it as "watching butterflies."

The first legislative considerations for the treatment of nonhumans had emerged in North America in the 1600s and 1700s, but the nineteenth century brought a

— 1809	Ephraim McDowell, firts successful ovariotomy
— 1822	Act to Prevent the Cruel and Improper Treatment of Cattle
— 1823	Royal Society of London reorganizes
— 1824	Founding of the Society for the Prevention of Cruelty to Animals (initially regarding carriage horses)
— Late 1820s	Burke and Hare murders
— 1831	British Association for the Advancement of Science
— 1831	Marshall Hall develops animal experimentation protocols
— 1835	Cruelty to Animals Act
— 1842	William E. Clarke, first tooth extraction under ether
— 1843	Reformed charter at the Royal College of Surgeons
— 1846	"Ether Day" in Massachusetts
— 1849	Cruelty to Animals Act (replaces 1835 Act)
— 1850	Royal Society of London gets grant-in-aid
— 1853	Queen Victoria gives birth to Leopold under ether
— 1863	Huxley becomes Hunterian Professor at the Royal College of Surgeons (Lawrence's former position), remains through 1869
— 1865	Bernard, *Introduction to the Study of Experimental Medicine*
— 1867	Joseph Lister *Antiseptic Principle of the Practice of Surgery*
— 1869–1870	Huxley is president of the British Association for the Advancement of Science
— 1870	Henry Lonsdale, *A Sketch of the Life and Writings of Robert Knox, the Anatomist*
— 1871	BAAS formalized Hall's protocols
— 1871	Friedrich Trendelenberg, first tracheotomy to accompany general anesthesia
— 1872	Lawson Tait, Intra-abdominal ligature for ovariectomies
— 1875	Cruelty to Animals Act (amends 1849 Act)
— 1875	Founding of the Victoria Street Society (December), founding of the Society for the Protection of Animals Liable to Vivisection
— 1876	Founding of the Physiological Society
— 1876	Royal Society of London's grant-in-aid significantly increased
— 1877	Anna Sewell, *Black Beauty*
— 1880	Lawson Tait, first abdominal appendectomy
— 1882	Founding of Association for the Advancement of Medicine by Research
— 1887–1888	Edward Berdoe's "St. Bernards" novel and follow-up volume
— **1893–1895**	**Huxley, *Evolution and Ethics***
— **1896**	**Wells, *The Island of Doctor Moreau***
— 1896	Opening of the National Antivivisection and Battersea Genral Hospital
— 1897	Founding of Our Dump Friends League (later Blue Cross)
— 1898	Founding of the British Union for Anti-Vivisection by Cobbe
— 1899	Lawson Tait, *Last Speech on Vivisection*
— 1900	Founding of the World League Against Vivisection (later the World League for the protection of Animals)
— 1902	Edward VII has burst appendix removed just prior to coronation
— 1903	Lizzy Lind af-Hageby and Leisa Katherine Schartau, *Eye-Witnesses/The Shambles of Science;* founding of the Antivivisection and Animal Defense Society; beginning of the Brown Dig Affair (1903–1910)
— 1906	Lin af-Hageby, *An Anti-Vivisection Declaration*
— 1907	Second Royal Commission on Vivisection, Brown Dog riots
— 1908	Founding of the Research Defense Society
— 1909	American Physiological Association develops animal experimentation protocols

Figure 2.3 Timeline for medical physiology, legal history, and anti-vivisection action.

significant jump in activism, especially in the United Kingdom. In 1809 the first such bill passed in the House of Lords, but was defeated in Commons, revealing a certain class issue in that bull-baiting was criticized but not fox-hunting, as was noted tartly at the time. The year 1822 saw the Cruelty to Cattle Act; the word was intended to mean "chattel," referring to all domestic animals, but its restricted interpretation sparked much confusion in application. All of this was occurring in a broader context of other significant social legislation, which is more than I can deal with here, including electoral reform, the ending of slavery in Britain in 1834, and the complex history of women and child labor. More than just legislation was involved, as compliance and enforcement are their own issue, and no policy could cope with the interplay among animal research, the use of executed individuals' corpses for medical studies, abusive treatment of women patients, and outright use of marginalized women as experimental subjects.

Further animal protective acts started consistently succeeding in Parliament, including the Cruelty to Animals Act in 1835, replaced with an Act of the same name in 1849. The issue gained a new focus when medical experimentation on animals boomed in the 1860s, hitting a climax in 1875, as the Anti-Vivisection Bill came before the legislature and a Royal Commission on Vivisection was appointed to investigate the issue. The debate was furious and wide-ranging, as everyone repositioned issues relative to one another and mobilized with every means. Ultimately the commission chose among these three:

- The Anti-Vivisection Bill, which despite its name was reformist rather than abolitionist, as supported by Cobbe
- An amendment to the existing Act, written by Darwin and Huxley
- A stronger amendment to the existing Act, which emerged in the Council sessions.

By this time in their fifties and sixties, respectively, well advanced in their careers, Huxley and Darwin had become power players in education and politics. Their activity overlapped considerably with the reform box in Figure 2.2, especially in developing public education curriculum and evening lectures for the working class. Regarding the Anti-Vivisection Bill, Huxley served on the Royal Commission and Darwin was deeply involved. At first glance, the Darwinian-reformer connection seems a pretty good match for policy alliance, as both men were noted in their dislike of the mistreatment of animals and in their straightforward support of the idea that human and nonhuman experience of life exhibited no observable differences.

However, it went very badly indeed. In the end, the Darwinians were summarily ejected from the issue, along with any and all thoughts of human as animal, leaving only the reformers and the medical researchers glaring at one another.

What went wrong? The history remains touchy enough for most accounts to be committed to blame of one sort or another, but a couple of points can be identified.

For Huxley's part, in some views, he simply began on the wrong side as a leading figure of the Royal College of Surgeons, as he initially managed physiologists' testimony in opposition to the Bill. He changed his mind after hearing some of it. Always an idealist for the ethical virtue of scientific thinking, he could not believe that men of science (a phrase dear to his heart) could be so cruel, until it was held under his nose. At this point, he dropped their unqualified support of the physiologists and cowrote with Darwin an alternate proposal, to amend the existing Act but not supporting the Bill.

For Darwin's part, the topic may have come to a head in a meeting with Cobbe, which would be a great moment for dramatizers of history, because they were a lot alike in some ways. Both were noted intellectuals who prized reasoned argument. Neither could be personally intimidated. Both loved animals and favored reform of some kind. Both combined hard-line ethics with a lack of sentimentality. Neither suffered fools. Neither was an extremist; Darwin did not support outright vivisection and Cobbe did not seek its abolition. They were also reasonably friendly acquaintances and had read one another's work. However, the talk did not go well, if later writings by both are any indication.

In the end, neither the Bill nor the Darwin-Huxley amendment prevailed, and what passed in August 1876, was a different amendment to the existing Cruelty to Animals Act. In comparison to the existing legislation, it was rather strong. It provided a list of standards and practices for using animals in research, and was explicit and informed regarding the techniques, including, for example, the accurate provision that curare alone was insufficiently humane, an insight that would not be applied to human medicine until the 1940s. What it didn't do was open research practices to public scrutiny or criminalize the mistreatment of animals used in research, focusing instead on liability for suit and fines for violations.[6]

As perhaps the perfect technical compromise among the competing voices, the new Act pleased no one, and it did not resolve the social issue so much as establish the vocabulary for more entrenched battle lines. Medical researchers of the day regarded it as plain interference and complied only under duress for generations, arguably hardening their viewpoint that they alone were the real heroes and protectors of society; medical students were to be available for violent counter-demonstrations and riots in later confrontations. The Society for the Protection of Animals Liable to Vivisection was founded in the course of the debate, and Cobbe founded the Victoria Street Society that December, with more severe views (later renamed the Anti-Vivisection Society).

The issues would come to another head in the first decade of the twentieth century with reports concerning the fate of a brown dog in a Royal College class demonstration. The resulting book and court cases provoked massive demonstrations, medical students carousing and rioting, round-the-clock police guard for a statue, and ultimately a second Royal Commission that further amended the Cruelty to Animals Act. As before, the identity politics hardened. Symbolically, the two windows were now fully in place, their discourse locked into a mutual set of defamations and uncompromising defiance, each fronted by a mob ready to throw rocks at the other.

Finally, then, here is Montgomery, the downside of the new professionalizing of science, a young man with some training but neither the intellectual drive nor the social ideals that give context to the scientific effort. He's the stereotypical useless student, already a famous image in the press well before the Brown Dog riots, not sure why he's there, more than willing to dive into camaraderie, relative freedom from authority, self-indulgence, and generally oafish excess. He doesn't represent one or more of the views so much as he falls between their cracks—a casualty of the times. Or rather, that was the case eleven years before the events of the story, and now he's a faded, bitter drunk.

This is also, I think, the moment when "Darwinist" as a term fell into its ultimate and most distorted form, not only as a pirate flag to wave, but as a target for invective. Cobbe and the general culture of research reform regarded the Darwin-Huxley role in the debate as a pusillanimous betrayal, legitimizing rather than banning research on animals, and her later writings reject Darwin's thoughts on human behavior as immoral. At least one strand within animal care activism would now characterize scientists as sadistic or pathologically insensitive. To support research but to acknowledge and regulate pain was derided as a false compromise by both sides.

Also, if evolutionary theory and the general viewpoint of natural history had any traction with medical researchers in the first place, it didn't any more. Don't miss the role of exceptionalism on both sides of this disaster. The reform viewpoint accepted "Darwinism" as a term on condition of full compliance, including the role of exalted humanity at nature's peak, and the medical research viewpoint rested firmly on the presumption that human welfare was not only the first priority, but a special one, not subject to question or compromise.[7]

So where did "Darwinism" go? To its ruthless and—in every sense of the word—*mean* version, always tagged with "survival of the fittest." This had been raised before as one meaning among many, and arguably the least intellectually grounded, but after this point it seems much more prevalent, especially when championed by Andrew Carnegie and John D. Rockefeller in the United States. It may be especially unfortunate in this context that the fifth edition of the *Origin*, published in 1869, included the "survival of the fittest" phrase, and if so, that's a grim legacy for the progressive Wallace, who'd certainly suggested its inclusion in a more optimistic spirit. What vanished in the tug of war were the genuine views of Darwin and Huxley themselves, among the very few genuinely criticizing human exceptionalism. From the 1880s forward, "Darwinist" would take its meaning from another kind of exceptionalism, which is to say, profiteering as virtue and racism in service to profit, privilege, and empire.

The Moment

So this was the cultural tipping point: what a mess! Biology had arrived as a nearly unified science, an educational curriculum, and an array of professions, but its most trenchant insights had proven too much. "Darwinism" as a constantly appropriated term had eclipsed thoughtful reading of Darwin's and others' work, so the rational discourse about human-animal research had deteriorated into juxtaposed fanaticisms, and society and science alike were recoiling from investigations of humans as animals. Without processing the human-animal question regarding usage and pain, medicine would proceed without built-in or useful principles for conducting research on nonhumans; without thinking of society and ethics as a human-animal feature, religion and morality would continue to be regarded as above or beyond human origins and animals' features. Worst of all, science and specifically evolutionary biology had acquired the status of a target for identity politics.[8]

So many high-minded and quite imaginary roles had emerged from these identities. Science was supposed to be a quest for absolute truth, social reform rallied toward an ultimate, "end of history" level of virtue and justice in society, medicine was upheld to provide an impossible degree of safety and comfort—all firmly rooted in the outlook that great and wonderful Man was taking on his noble role and purpose at last. Talking about humans as animals without caricature throws these visions into disarray, and no one knew what to do with that.

In such heady, self-congratulatory context, how is a society to resolve the naturalist versus medical approach to biology, the church versus science tug of war over education, the tensions among science, technology, and power, the clashing values among animal research versus regulation versus anti-vivisectionism? The answer is, as long as everyone is mired in human exceptionalism, it doesn't. And it hasn't.

Fortunately we have a novel that didn't miss the boat: fully informed and fully provocative, precisely capturing the cultural conundrum as it settled into a knot and as it has remained. You will not find in *The Island of Doctor Moreau* any Darwin-*ism*. That absence is startling, given the history I've presented here. Whether Wells "meant" to do it or not, I have no idea, but put simply, the story doesn't *care* about appropriating Darwin's ideas as an "-ism," and concerns itself—uniquely—with the hammer blow delivered by what Linnaeus, Lawrence, Darwin, and especially Huxley actually said. It is rooted directly in *Evolution and Ethics* and its underlying confounding factor of human exceptionalism, to challenge the dichotomy of Man and Beast, and asks, when this dichotomy is removed, what might we see. It is also rooted directly in the two-windows crisis, simultaneously gruesome and stern in its critique of that entire construct. In a sea of social and political views that never violated the Man/Beast taboo, only Wells had found the way.

Readings

Thomas Huxley's *Collected Essays* from his later life are all recommended reading. I can't recommend actually reading Linnaeus's *Systema Naturae*, but David A. Baum and Stacey D. Smith, *Tree Thinking* (2012), provides an accessible introduction to modern systematics, and Patricia Fara, *Sex, Botany, and Empire* (2004), shows how Linnaeus's ideas became widespread. For the Linnaeus-Gmelin dialogue, see Tore Frängsmyr, Sten Lindroth, Gunnar Eriksson, and Gunnar Broberg, *Linnaeus, the Man and His Work* (1983). It's also unfair to leave out an earlier writer, the irrepressible Julian Offray la Mettrie, whose essays from the 1740s should be more widely read: *The Natural History of the Soul, Man a Machine* (1994), and *Man a Plant* (1994).

The standout modern reference for these issues and this period is Adrian Desmond, *The Politics of Evolution* (1994), an account of the rough-and-tumble social strife in which the scientific terms took new shapes. Helpful work on Spencer's ideas can be found in John Offer (editor), *Herbert Spencer: Critical Assessments* (2000), especially Volume II, *Spencer, Darwin, and Social Darwinism*.

Overviews of anti-vivisection and evolutionary theory during the nineteenth century include Richard D. French, *Antivivisection and Medical Science in Victorian Society* (1975); Susan Hamilton (editor), *Animal Welfare and Antivivisection, 1870–1910* (2004); and especially Coral Lansbury, *The Old Brown Dog* (1985), which provides a very useful overview of events and values leading to the Brown Dog riots of 1907.

Many nineteenth-century anti-vivisectionist writings are worth modern consideration, including Francis Power Cobbe, *Illustrations of Vivisection* (1887) and *Darwinism in Morals and Other Essays* (1872), and William Youatt, *The Obligation and Extent of Humanity to Brutes* (1839). Edward Payson Evans, *Animal Trials* (1906), reveals another remarkable aspect to the issue, when nonhuman animals are prosecuted as responsible perpetrators of human-defined crimes. See also the US zoologist J. Howard Moore, *The Universal Kinship* (1906) and *The New Ethics* (1907), similar to Huxley in speaking against the overwhelming cultural trends.

Critical and complex historical accounts from animal welfare scholars and advocates are found in Rod Preece, especially *Animals and Nature* (1999) and *Brute Souls, Happy Beasts, and Evolution* (2006), and Peter Singer, *The Expanding Circle* (2011). Both authors' considerable bibliographies are recommended. The accounts of Darwin's and Huxley's roles in policy problems require a certain broad reading from different times and multiple viewpoints; see Rod Preece, *Animals and Nature*; William Irvine, *Apes, Angels, and Victorians* (1983), and especially James Rachels, *Created from Animals* (1990).

Notes

1. Linnaeus's work is genuinely seminal to biology as it developed, immediately and to the present. Lawrence explicitly supports his methods and the implications of a branching history of life rather than a stepwise, linear process. The *Vestiges* also incorporates his points in detail. Darwin's ideas synthesize explicit evolution from Lamarck, Linnaean taxonomy, and causal thinking. By contrast, in ignoring Linnaeus, Spencer was outside the intellectual community of biologists from the very start.

2. Original letters exchange 1747, cited in *The Linnaean Correspondence*, http://linnaeus.c18. net/Letter/L0783. The Linnaeus Correspondence site is provided by the Royal Swedish Academy of Sciences, Uppsala University, and the Linnaean Society of London. The translator is not credited.

3. The period from 1900 through the mid-1920s may be considered the "closer" for biology as a modern science, with the decreased emphasis on Haeckel's concept of "ontogeny recapitulates phylogeny" and the Hardy-Weinberg equilibrium concept resolving the perceived conflict between Mendel's and Darwin's ideas. However, I think all the questions answered during this time arose during or before the 1890s, and I'm more interested in the questions being articulated than in the answers.

4. I do not include "Social Darwinism" in Figure 2.2 because it's a retroactive term. The phrase originated with Joseph Fisher as described in *The History of Landholding in Ireland* (2008), but was very rarely used, and only as an accusation, not as an adopted viewpoint. It was also used by Oscar Schmidt in 1877 as an accurate descriptor for the Revolutionary view I've included in my diagram, but that usage has not survived. The term's well-entrenched usage today is best placed with Richard Hofstadter's *Social Darwinism in American Thought* (1944), and its application to people like Francis Galton or Herbert Spencer is a retrospective judgment rather than a representation of a unifying view. As far as I can tell, during his lifetime, Spencer's writings were more frequently cited by progressives. In my construction, the view that Hofstadter describes would be a mash-up of the Imperial and Darwinian views, seizing upon the "survival of the fittest" phrasing, regardless of what Spencer did or did not intend.

5. I have not described the Revolution and Imperial views in detail, but by the 1890s, they were so prevalent and had such a high impact as literally to define global politics through today. The latter half of the nineteenth century brought wars of empire, including India, Afghanistan, and the Boer war; the US civil war; the unification of Germany and its *Blut und Stahl* ideology; astonishing misery in urban areas; and the appearance of Fenians, Zionism, radical Islam including the renewal of Wahhabism, and revolutionary communism, among other movements.

6. Rod Preece provides a harsher assessment of Darwin's role in the negotiations of 1875 in his "Darwinism, Christianity, and the Great Vivisection Debate" (*Journal of the History of Ideas* 64(3): 399–419, 2003) and in *Awe for the Tiger, Love for the Lamb* (2003).

7. Contradicting my neat two-windows story, the activist most instrumental in the Brown Dog affair, Lizzy Lind af-Hageby, cited Darwinism to challenge anthropocentrism, in contrast with Cobbe's legal and moral conflict with Darwin.

8. In the United States, evolution was initially not a hot-button issue on a national or even state scale. Evangelical Christian organizations and churches did not consistently target Darwinism for vilification, that is, so-called Creationism did not become politically volatile until after about 1900; until then, they were vocally for or against on a piecemeal basis.

3

Don't Meddle

What is the science in *The Island of Doctor Moreau*? Is the book critical of science? Of course it is. The question is which meaning of "critical" is involved: investigation or condemnation?

It could add up to a thriller plot: crazy science produces abominations, with the subset of cruelty to animals thrown in for extra moral censure. Everyone is familiar with the default science fiction story in film, in which scientific experimentation proceeds without regard for morality or consequences, people suffer due to the outcome, heroes (who know it should never have been done) struggle to survive in difficult circumstances, the results are uncontrollable and dangerous to others, something happens that is disgusting or spectacular or both, and the meddling scientist is punished. The scientist has as well one or another social disorder wrapped up in his motivations, in some combination of pure knowledge, the greater good, intense idealism, wounded ego, and viciousness, which focuses his attention to the point of passive or active cruelty. That's the assumed and expected plot: bad scientist, don't meddle!

I can understand why a person might think that "don't meddle" is the point that the novel is making. These issues are invoked in the story, in fact in so many words, but each detail—the crazy scientist, the violation of natural law, the ensuing disaster or retribution—turns out to be reversed in some way, or off-message enough to throw the whole model into question. Unfortunately, as with *Frankenstein*, expecting this plot has a way of subverting the experience of reading, causing the fiction itself to be missed.

Moreau is terrifying in his determination, but he's not barking mad and has no interest whatever in defying, outdoing, or being God. He's actually unswervingly religious in standard late nineteenth-century intellectual terms. His fixed idea (monomaniacal, to be technical) lies only in his adherence to a higher state for humans as distinct from nonhumans, which is to say, what most people believe anyway, which puts a rather different light on his "mad" science. He isn't the main character, even by implication, and is killed early relative to the in-story chronology and its climactic events.

His project is not limited to an engineering or invention project, "do it just to do it," but he is investigating a valid theoretical point: whether the ordinary processes of the body as observed in nonhuman animals are capable of functioning as a human being. He's not seeking to improve upon human beings, or to emulate evolution as such, but to evaluate, as he correctly calls it, "the plasticity of the flesh."

Its design isn't so good, lacking controls and a defined performance variable, but the techniques work: the nonhuman animals are transformed successfully. I know this is a bold claim, to be developed in Part III of this book in detail, but for now, consider that the Beast People's appearance is disturbing but not outrageous, with one exception; they are not covered with fur and do not have protruding fangs or animal heads, and everyone who meets them, even up close and unclothed, thinks they are people. Their actions are interpretable and responsive, not random uncontrolled urges. Most of the "beastly" behavior in the story is evidenced by the people, not the Beast Folk.

There's a lab accident, but it poses no general, society-shaking danger, and is better understood as a personal act rather than an accident as such. At the end, the Beast Folks' regression is not a vengeance-driven outbreak of frenzied bloodshed and doesn't deliver retribution to anyone.

Something's missing from the "don't meddle" plot—where's the danger-filled disaster? Where's the retribution? Where's the lesson? They're not there.

Movies and Moreau

The subversion of the novel *Frankenstein* into the "don't meddle" story is at least partly understood, and coming under more and more scrutiny. However, the same subversion of *The Island of Doctor Moreau* remains culturally unrecognized, so that the novel's content is frequently misrepresented as this exact anti-science story. Here, I aim to outline and compare the issues of abomination and other specific plot points in the films to show what the novel is not. Seven versions have been made.

- 1913: *Ile d'Epouvante* (*The Island of Terror*)—I'm still looking for this one, but sadly, it might be a lost film, like the first film versions of *Frankenstein* in 1911 and 1915. It was produced in France by the Société Générale des Cinématographes Éclipse and was directed by Joë Hamman (not to be confused with the 1970 Italian film of the same title produced and directed by Mario Bava, which is not based on Moreau).
- 1932, released 1933: *Island of Lost Souls*—this is a Paramount film directed by Erle C. Renton, with Charles Laughton as Moreau. It's played the same role as the 1931 *Frankenstein* in establishing the story's content in popular culture, even more completely if that's possible.
- 1959: *Terror Is a Man*—this is a Hemisphere Entertainment film directed by Gerardo de Leon (billed as Gerry de Leon), with Francis Lederer as Dr. Charles Girard, produced in the Philippines by the famous filmmaker Eddie Romero.
- 1972: *The Twilight People*—this is a Dimension Pictures (Four Associates Ltd.) film, the first version of this story in color, also produced in the Philippines, produced and directed by the same Eddie Romero, with Charles Macaulay as Dr. Gordon.
- 1977: *The Island of Doctor Moreau*—this is an MGM film directed by Don Taylor, with Burt Lancaster as Moreau (first name Paul).

- 1996: *The Island of Doctor Moreau*—this is a New Line Cinema film directed by John Frankenheimer, with Marlon Brando as Moreau, famous for quirks arising from its disorganized production history as described in the documentary *Lost Soul* (2014).
- 2004: *Dr. Moreau's House of Pain*—This is a Full Moon Entertainment film directed by Charles Band, with Jacob Witkind as Moreau. It is definitely not a mainstream item and did not receive a theatrical release.

Across the films, the fundamentals of the plot differ from the novel in consistent ways: Moreau, by whatever name, makes monstrosities and the monstrosities go nuts, serving as both rampaging monster and retributive villagers at once, and kill him as he richly deserves. The details that deviate from the novel and make this plot into "don't meddle" are quite consistent:

- The project's goal is typically more grandiose, including improving upon humanity rather than arriving at it, outdoing or overcoming evolution, and in at least one case, eradicating evil.
- Beyond straightforward questions of ethics, something is psychologically very wrong with Moreau, in some combination of hubris, sexual perversion, active or inadvertent sadism, bullying, and impulsiveness in his experiments; there's always a kick-the-dog moment or revelation of past wrongdoing at the very least.
  - His demeanor varies, but always along the mad-scientist scale, whether psychopathic coldness or dysfunctional paternalism.
  - He often turns toward Prendick as a research subject.
- The work is explicitly called abominable in various phrasings, which to be fair, happens in the novel, too, but here it's a simple identifier that I infer to be authoritative and to which Moreau has no meaningful reply.
  - The subjects' agony during their transformation is downplayed from the novel, and their unjust treatment afterward is ramped up; in the three mainstream films, the Beast People are exploited and miserable, treated as a colonial village and subject labor force.
- They are physically obviously nonhuman, especially in the later films, when they are effectively bipedal versions of the various animals.
  - Related to the visual effect, they speak in movie Tarzan-talk, with limited vocabulary and broken grammar.
  - They are often baffled by simple tasks or are easily fooled, similar to "natives" in movies with racist and colonial tropes.
  - They experience inexplicable rages and surges of violence; apparently their native animalism involves injuring creatures around them for no reason; in

the 1977 film, they kill and presumably eat Braddock's companion in the first moments of the story.

- There are no ordinary, run-of-the-mill female Beast People; either they are absent or one is present who looks very much like a human.
  - The concept of "animal" is synonymized with rapist in at least two of the films.
  - In the 1996 film, some female Beast People are present in addition to the single human-looking one, but are extremely incidental.
- The Law is invented and imposed by Moreau, and it is nothing but another form of pure cruelty, arguably not perceived much beyond pure programming by most of the Beast People.
  - Some do express dismay at the hypocrisy behind the Law.
  - They are miserable in their lack of natural identity, including self-recognition as "Things! Things!" as expressed in *Island of Lost Souls*, or "Father, what am I?" in the 1996 film, similar to the "We belong dead" line in *Bride of Frankenstein* (1935).
- The taste of blood prompts the rebellion, which results in the vengeful rampage at the end.
  - The fully unleashed "animal" is a raving berserker, corresponding to the term being employed as derogatory, so the ending rampage is an orgy of frenzied vandalism.
  - This can be a bit tricky, too, because the rampage is "bad" but the vengeance is "good" when it targets Moreau.

Since this story is about Moreau, Prendick becomes a bit superfluous, which is probably why he has such a different personality in each film. He's never the scientifically informed, moderate progressive, as in the novel. In the earlier films, he's a hard-assed engineer, although with a streak of romance, upgraded in the Philippine versions to a genuine tough guy. In the later films, he's generally an ineffective character. Moreau's assistant Montgomery turns out to be more important, with extremely different characterizations and plot roles.

I don't want to give the impression that I hate the films. I freely admit that I own them all. Each does successfully generate a gray area for the moral capabilities of one or more of the Beast Folk, usually limited to the parameters established for Frankenstein's creation in the 1931 film: only human enough to further establish the cruelty and irresponsibility of the scientist. Some of the performances and ideas toward that gray area are remarkably good. Movies are irresistible discussion topics, so throughout this book I'll point to them when it seems helpful or interesting to the current topic, but these discussions will appear in boxes to keep the movie talk from overwhelming things.

In *Island of Lost Souls* (1932), science is torture, and it won't work. Moreau is openly sadistic and more than a bit of a pervert, especially creepy because throughout the first half of the movie, the content of his dialogue is picture-perfect intellectual, but delivered with the mannerisms of a criminally mis-behaving schoolboy. These get more and more intense until the net effect is pathological hypocrisy: the heights of averred intellectual achievement motivated by sadism, megalomania, and a nasty smugness in getting away with it. The progression hits its peak at the precise halfway mark of the film, when Moreau first lounges on one of his own operating tables like a Roman sybarite, then gloats over his "failures" slaving away on the pedal-operated generator, and caps it with self-congratulatory blasphemy, in his "Do you know what it means to feel like God?" line. A big part of the plot is manip-ulating the hero into a triangle between his fiancée and Lota the Panther Woman, and in connection with this, several of the Beast Folk are interest-ing characters and very well acted, making their drama more interesting than Moreau's fate.

Terror Is a Man (1959) focuses on the romantic triangle among the hero, here named Fitzgerald, the single Beast Person (a male panther), and Dr. Girard's estranged wife, and it's the single Moreau film that stays with the castaway's personal drama rather than on the researcher's fate. Girard is also unique in film versions of Moreau because he's completely sane, as the hero acknowledges without sarcasm, only matching the "don't meddle" scientist profile in his callousness regarding others' safety and lives. The research is apparently going poorly in that the subject repeatedly escapes the lab to wreak havoc on the islanders, but the later part of the story steadily improves the panther man's moral standing over the other characters, even the hero. Unlike all the other films, he becomes (slightly) less savage as the story pro-ceeds, rather than more, and Girard's techniques are at least implied to be sound. It's from an era of slower paced filmmaking that doesn't translate well today, but I like its seriousness and find the panther man's fate moving. It also features the cheesiest promo gimmick: the soundtrack rings a bell prior to the surgery scene in order to warn the "squeamish and weak of heart" not to look, which was advertised with relish.

The Twilight People (1972) is a fine example of a grindhouse film, and like many of them, it includes some good ideas and moments in all the jaw-dropping excess and catch-as-catch-can special effects. However, the experiments and the quite hideous Beast People are not very important to the story, except as spectacle. The main characters are all humans, and their conflicts are again a romantic tangle, so the research subjects are not there for much more than mayhem. The main villain is the assistant, here a sadis-tic mercenary, and the main problem is solved when the hero marshals the

Beast People to slaughter him and his henchmen. Even Dr. Gordon isn't very engaged in the problems of the plot and comes off as numbed and exhausted, although this is livened up a little because he meets his end at the hands of his wife, one of his subjects.

In *The Island of Doctor Moreau* (1977), Moreau is initially harsh and cold, sinister in his intellectualism, but is then abruptly revealed to be deranged by fear of his own creations, including shouting at and whipping them for no conceivable reason, as well as the classic turn of shooting his own minion when he gets too assertive. The story includes the 1932 version's plot point of setting up the hero for sex with the beast-woman, this time successfully (it's the 70s now, after all). Here the mirror-effect of beast-to-human versus human-to-beast is at least mentioned, but significantly, the reverse procedure doesn't work. Also typical of 1970s cinema, the content is spectacular but scattered and contradictory, including Moreau's odd Christ-like posture in death.

In *The Island of Doctor Moreau* (1996), Moreau is a cultist-paternalist, deranged in that he is oblivious to the grotesque misery of his creations. His personal style aside, Moreau's plot role is almost identical to that in *Island of Lost Souls*, as the research is straightforwardly unsuccessful, the basic dysfunctional relationship among the humans and transformed animals breaks down, and the former get what's coming to them as the rebellion becomes a rampage. It lacks the romantic subplot entirely, therefore losing intriguing aspects from the earlier films, and the narrator has no agency throughout the story. It may be a mess, but it shouldn't be dismissed. During its complicated production, someone took a few ideas seriously, especially about religion as a form of social control; it's also clearly influenced by *The Twilight People* (among other things, the slight name change from Ayessa to Aissa) and Brian Aldiss's novel *Moreau's Other Island/An Island Called Moreau* (the modern context of political strife, the rifle and guerrilla revolution, and some other military implications); and someone misread the ending of the novel at least enough to reference it.

Doctor Moreau's House of Pain (2004) comes from the same tradition and aesthetic as 1970s grindhouse, although its venue is home viewing rather than exploitation-film theaters and drive-ins. It is explicit and excessive. If you like this sort of thing, this film definitely fits the bill, and if you don't, it will cement that opinion into place for good. All that said, Moreau's goal is one of the more similar to the textual version in film, given a twist toward the ruthless insofar as human organs constitute part of his techniques, and in considering the hero as mating stock. It's the only film in which Moreau is not killed by his creations. The story includes thoughtful conceptual punches, including the creations holding Moreau prisoner and demanding further work upon themselves so they can join human society. Although the experiment has "gone out of control," it does so toward the end of wondering whether it has actually sort of succeeded.

In Carol Clover's *Men, Women, and Chainsaws*, she suggests that lower-budget films made outside mainstream distribution typically address disturbing content more directly and with more thematic follow-through in their endings, and should not be dismissed due to their excess. I find her case to apply well here, especially regarding the 1959 and 2004 versions.

The term "manimal" is absent from the novel and all the films except for *Dr. Moreau's House of Pain*, although I often see it misattributed in collective references to the novel and its adaptations. It may have come from the 1980s TV show *Manimal*, which has nothing to do with Moreau, or from the US rock band the Manimals, which uses many motifs from *Island of Lost Souls*.

Ultimately, the issue boils down to one question: Is the science frightening because it violates natural law, or because it shows that something we believe is not as rock solid as we thought?

Sorcery

Movies didn't invent the "don't meddle" story, whose formal origin was probably *Presumption*, Peake's adaptation of Frankenstein, in the early 1820s. This is the version that introduced the corpse or near-corpse automaton concept for the creation, as well as the comedic element of Fritz, better known as Igor.[1] Apparently there was quite a genre of theatrical versions in the late 1800s. Movies certainly embraced the story immediately, with two versions of *Frankenstein* and one of *The Island of Doctor Moreau* in the early 1900s, almost as soon as commercial film-going was invented. The defining longer-feature version came with Dr. Rotwang in *Metropolis* (1927), which introduced the mad scientist's connection with ruthless industrialism, but the *ur*-example is the Universal Studios film *Frankenstein* (1931), which set the template for so many, many movies afterward. Not only is the creation a walking corpse, constructed from stitched-together bodies, it has the brain of a psychopath; although it's the tiniest bit sympathetic in receiving only abuse, its rampage is a foregone conclusion, and the villagers' revenge on Frankenstein is nothing but justice. Frankenstein himself takes the theatrical version's hubris to the point of pure mania.

In these stories, science as a process bears no resemblance to reality, and I'm not talking about picky details, but its absolute identity. The debate among members of a community, the professional context of journal publishing—all are absent. Research is funded either out of one's pocket or is institutionalized at the extreme forms of privatized product or government conspiracy. It's all engineering research, as the only activity seems to be invention, that is, science isn't a process at all, only product. Similarly, scientific results do not yield insight or contribute a nugget to a more complex debate, but rather boil down to this widget or project either working or not working. Everything else is mere technobabble, accurate or otherwise, and always gaudy with dangerous details. "Testing" doesn't mean testing at all. Ooh, glowing, smoky, genetic stuff! See what happens! Bonus points for trying it on yourself.[2]

Movie scientists as people are a little closer to reality. Real scientists' education retains many features of the medieval guild, defined by mentor-journeyman relationships and patronage, and this leads to the content seeming arcane to everyone else. Most scientists aren't employed like normal people; typically they work for one institution, like a university, but their research is funded by another, like a federal agency, and their work is assessed by yet a third, the journal-publishing mechanisms of club-like scientific organizations. It's genuinely impossible to explain to people "what you do" in terms of activities and payment, and why one both does and does not have a boss or work colleagues in the usual sense of these terms. The work itself is both public, funded mostly from the federal budget and technically accessible to anyone through publication, and private to the point of being cloistered, being professionally invisible and intellectually obscured. They're strange beasts politically, too, considering the role of scientists in high-profile technology like the Manhattan Project and the spy fever that swiftly followed, and the vocal presence of scientists in activism—including militant activism—in later decades, for example in Earth First! The significant yet mysterious social role for the scientist that prevailed during the writing of *Frankenstein* persists today on a larger scale. They are both insiders and outsiders, the servants of untrustworthy power and the untrustworthy nonconformists subverting or threatening the normal or reliable status quo. Fear aside, these perceptions are quite accurate—science simply doesn't fit into the familiar frameworks of employment-based identity and political loyalty.

Two stereotypes emerge, reflecting the uneasy, mixed perception. One is the loopy, speculative geek, associated either with a university or a very cloistered government program, and the other is more of a company man, with industry or with some practical branch of government. The latter is an important historical construct, as it was deliberately built and reinforced during the early Cold War, especially from the writings of C. P. Snow—the scientist as a patriotic, clear-eyed visionary of the future, with aggressive engineering being the world's only hope.[3]

Moreau does not indulge in one of the fictional scientist's famous habits, self-experimentation, but it's more common in reality than one might think. Some of it was inadvertent, as when organic chemists tasted their concoctions as a means of analysis until modern instrumentation became available. Some of it was deliberate and as loony as anything on film. In the time the novel was written, Max Josef von Pettenkofer drank cholera bacteria in 1882 and Charles-Édouard Brown-Séquard injected himself with testosterone in 1889. More recent examples include Albert Hoffmann self-testing with LSD in the 1940s, and Barry Marshall drinking *Heliobacter pyloris*, as published in 1985.

The "don't meddle" story turns all these things up to the point of pathology. The scientist's professional isolation often includes physical isolation, in turn identified with his or her psychology: at best socially awkward, whether eccentric to genuinely nuts, painfully friendly to painfully rude, or some combination of endearing, sinister, and pathetic. At worst, this grades into sociopathy: abusive treatment of animals and assistant, dismissal of danger to the public, and open defiance toward such considerations. He or she is always curious regardless of consequences, sometimes to the point of being too dumb to live, with weird blind spots, such as a curious infusion of fact-faith dogmatism into the dizzy specula-tion. For someone whose work effectively consists of inventing something new and pushing the "go" button, he or she is strangely unprepared for surprises.[4]

If these observations were merely whining about failed depictions, scientists could join the long list of misrepresented film professions, which is sociologically interesting but probably not very important. However, these details factor into something odd about the "don't meddle" plot that never changes. Taken literally, its various criticisms of science tend to cancel out. If it's not cruel, is it still wrong? Or if the scientist isn't a weirdo? Or if it's not misused? Or if there are no accidents? If it produces the intended results? Since these components are usually blended, it's impossible to tell what's actually wrong with it, aside from being a destructive or disgusting spectacle. Where's the moral failing: in the science itself, in human applications, or in the dangers of the procedures?

For example, look at that so-common accident, with its subset of unantici-pated results or misuse by somebody. Notice that the scientists must be distinctly morally blind in conceiving and carrying out the project, because the designated sane character in the story instantly sees its inherent wrongness. To a decent person, the project is *obviously* wrong, against Nature, and he or she knows that this wrongness must "out." Therefore the accident is not a mere procedural risk that could happen to anyone, but a retributive backlash. Even if the scientist were to run the procedures exactly right, harm no one, hurt no creature, spill noth-ing, indeed cover every imaginable base, the universe will come along with its coincidences and punish him for doing it, with collateral damage guaranteed. If it's not already obvious how the disaster arrives, then it will anyway, because it *must*—Nature itself, offended, accepts no less.[5]

This principle is especially obvious in *Jurassic Park*, whose plot is a perfect blend of *Frankenstein* (1931) and *Island of Lost Souls* (1932), and in which the scientist is especially feckless in his naiveté. Malcolm (Jeff Goldblum) is presented as the mathematician and the allegedly objective scientist, but he is scripted and played as a clergyman in all but name, who flat-out says that God's plan for the dino-saurs was extinction, so therefore the scientists should have known not to resur-rect them. Not only is such an act evidently wrong, but it cannot work—even its brief appearance must yield a corresponding catastrophe, through the operation of literal natural law.

I rather appreciate the film's honesty in this regard: it's a spiritual criticism, about the intentions of the universe and our proper place in it, and has nothing to do with danger, ethics, or motivations. The project doesn't merely undergo a snag, but was intrinsically blasphemous from the start and therefore must result in backlash, as the universe responds morally. I stress this point: the "don't meddle" story is about dealing with the abominable. It isn't about science at all, but sorcery. It's rooted in and relies on absolute fear.

Yes, it's trouble. Scientific work is not socially or morally neutral. It constantly provokes moral and political confusion, both personally and regarding institutions of power. But none of that is being substantively addressed as long as people talk about "uncontrollable forces" and "things man was not meant to know."

Scientific thinking is an ordinary, easy to understand human activity. We all interact with the physical world all the time, and some of that interaction involves trying to figure it out. If you tinker with any of it and consider how whatever you did interacted with whatever it is, then you're being scientific—even more so if you're thinking about general or underlying principles ("how plants grow"), rather than merely your current bit ("my tomatoes"). Scientific thinking is so normal that people forget to recognize it when it's not dressed up in professional trappings and technology.

Science means more than the thinking and reasoning process. In a given society and period, sometimes—as now—thinking this way is formalized into institutions, professions, and ways to assess and record it. It's about the topics of study, how they've changed, what's been produced, what conclusions and general principles have become ordinary knowledge, and what technology is under way. Referring to a *scientist* is talking about a social status, a set of educational requirements, a range of jobs, means of funding, and many other embedded features. Modern scientists cannot help but have defined, albeit squishy, relationships with other societal institutions and roles, like education, commerce, and manufacturing.

At first glance, their role may seem to be simply contributive. However, science *is* genuinely disruptive. Its output consistently changes two things: the concrete technical nuts and bolts of living, and the explanations for everything about living. It's as if you took art or writing of any kind, with all of its individual, unpredictable political potential, and maximized its direct impact on people's lives. It seems all bound up in societal institutions and social norms, but its emergent effects cannot help but sooner or later turn around and challenge them.

The most obvious and relevant case is technology, which is a constant side effect of scientific thinking, and which gets into society and affects people there (Figure 3.1). You can see three kinds of science at work. Basic research is sometimes called "idea-based," because a given project typically examines a theoretical

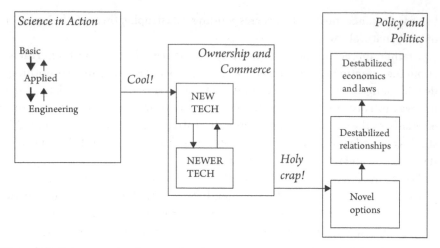

Figure 3.1 Science, technology, and society.

point. Basic researchers continually comb historical science for claims that don't make sense or for claims that might, but were overlooked. Many of them collect information about areas and types of creatures that haven't been studied yet.

Applied research is carried out in the context of an acknowledged human problem, often medical but also ecological, and always economic. Not enough is understood to start working on a direct solution, so anything and everything that might be relevant gets studied in order to generate a better understanding of the background of the problem. Applied researchers' work can sometimes take them a long way from the immediate problem, because contrasting species or systems, or a counter-example like a desert instead of a swamp, can contribute immensely to the discussion. They also draw heavily from existing basic science, often putting it in a new perspective.

Engineering research is what non-scientists understand best: you invent a pill, a widget, or any other similarly understandable object, often for sale. The examples fill the popular understanding of science: light bulbs, powered flight, firearms mechanisms, industrial agriculture, the Manhattan Project, space flight, the birth control pill, the Internet, and so on and on. As with applied work, engineering researchers draw opportunistically from the whole range of existing scientific literature, and often a given project emerges from prior work with no make-a-device content at all. Their work is usually carried out in terms of ownership, state policy, marketing, and commerce.

These distinctions show how unconstructed science is. No one does the whole thing! A given researcher does not work on the entire trajectory of basic to applied to engineering, fueled by an individual vision. Such trajectories are typically identified only in retrospect, across many different scientists, often across generations, and in the context of a spirited debate about every step and idea. You might find the best work was done by someone whose immediate conclusion was wrong. You can find "skips" of different kinds, such as when a given idea was considered loony or insignificant until someone went to the trouble to rehabilitate it (such as the catastrophic demise of the dinosaurs), or when a project's whole reason for being gets revised retroactively in the face of new evidence (such as the Human Genome Project). Even more important, no one can anticipate what basic information turns out to be most relevant to applied projects, or from applied projects to engineering ones. Each level or type is a big fermenting intellectual mess, with people scooping old and new stuff out of there all the time.[6]

One might think that technology always comes from engineering research, but in fact, most scientists are not inventors in that sense. Instead, they jury-rig a lot of imaginative practical devices as some means to get the research done. Many of these then effectively become inventions, often retooled to different purposes. So it doesn't matter whether people are trying to invent things; if science is happening, then new technology appears more or less steadily. This is why the technology arrow comes out of that whole box, not specifically out of the engineering category.

When technology gets out of science and into the hands of commercial owners, it undergoes more changes, for instance, when a competing company puts out its own version of a new widget, just different enough not to violate the patent, and then the first company upgrades its original version, and they go back and forth like that.

When people in the larger culture make contact with the new technology, by definition, it alters their options—often very important options. Think of anything we already do, and consider doing it more easily than you did before, or without worrying about this other thing happening, or through some alternative that you could not do before at all, and so on. Not only is the effect personal to you, but it changes everything about the activity, such as how many other people it can affect, who can and cannot use it, how cheap or expensive this particular way to do it is, and its indirect effects on you or on the environment. In this case, the technology "attacks" social relationships, altering the physical dynamics that everyone had been using as a baseline for their positions. It doesn't merely introduce a problem or solution to an existing situation, it changes the whole situation.

Consider the birth control pill, which first appeared as a consumer product in 1960 as the Pill (capitalized), referring to the product Enovid in the United States. It was invented and produced in a mash-up of policy issues, some obsolete and some still with us, including sex education, women's rights in general including abortion (itself soon to undergo legal review), competing birth-control technologies, population control, and the general relationship of the Roman Catholic Church to US politics. In *America and the Pill*, Elaine Tyler May describes the idealistic or horrific proposals that people raised concerning the existence of an easy-to-use, reliable, and female-controlled contraceptive technology. Few of these came to pass as hoped or feared, and more to my point, what did in fact happen was *not* in that mix of debates: perhaps unbelievably to modern eyes, no one anticipated that having the Pill around meant that a lot of women would want to have a lot more sex.

May cites the 2004 report from the US Department of Health and Human Services: 98% of women who engage in heterosexual activity to any extent use contraceptives, 82% being oral contraceptives, 79% including their first sexual experiences. This is a big change from 1960! Sex is different. Love is different. Careers are different. Being a "woman" is different. Being a mother is different. Parenting is different. Even not using it becomes a variable in the context of its existence. The new standards and practices weren't and probably could not have been predicted, but they are the new parameters of many people's life experience. This is about values, because new technology changes the game so much that the social discourse of right and wrong is literally thrown into a new environment.[7]

Consider the rate of such changes, across technologies and throughout society. Historically, modern institutional science coalesced in the late decades of the 1800s, with an additional boost in the 1940s and 1950s—squarely in the context of the Industrial Revolution, the world population explosion, the rise of the modern state, and profound changes in labor types, education at all levels, and urban living. Science became professionalized, an option for middle-class wage earning and specialized vocational training. It also became embedded in distinct forms of institutional memory, university accreditation, and state funding. In this context, the ongoing production of new technology is institutionalized, commercialized, and mobilized in a constant state of accelerating fast-forward, and with broader and more significant impact.

In 1970, the journalist Alvin Toffler published a popular book called *Future Shock*. Its specific content seems quaint to me now, but the underlying concept is valuable: that historically, new technologies were culturally absorbed through a certain buffer of time, but now, the rate and breadth of introduced technologies will overwhelm the adjustment process altogether, throwing people into a state of constant values-violation and confusion. His model is too rosy about past historical adjustments being so easy or positive; still, allowing for some sensationalism about how scary (or cool) this is, and recalling that our current population and industrialization levels are historically unique, I think some degree of future shock is with us today, and that it provides fertile ground for the sorcery reaction to become an embedded, constant feature of modern society.

Life Science

What makes biotechnology different? Arguably, not very much at all. From the UN Convention on Biodiversity in 1992, biotechnology is defined as

[a]ny technological application that uses biological systems, living organisms, or derivatives thereof, to make or modify products or processes for specific use.[8]

This definition applies in full to the earliest forms of agriculture (gardening), brewing, food preparation, first aid, and domestication of animals. It applies to the Pill as just described. In fact, all medical technology is biotechnology and always has been.

The issue is therefore not *bio*technology, but bio*technology*, meaning what we could do with it *just yesterday*, which kicks off the reactive, frightened response *right now*. Let's take cloning. Imagine for a moment everything you ever heard or knew about cloning. What's the big deal?

Cloning can be done in several ways, but I'll focus on the technique called somatic cell nuclear transfer, or SCINT. Here are two examples, one fanciful and one real.

Let's take me: a somewhat healthy middle-aged adult human male. Lots and lots of living cells compose my body, and I won't notice if you take some. However, I am not an embryo, and my cells aren't going to act like embryonic cells very easily. You should go after some of my epithelial tissue, which is at least a bit similar because those cells are constantly dying and being replaced, so a lot of ready-to-use but not yet active cells are waiting their turns. Grab one of these and put it in a jar.

Meanwhile, two individuals whom we shall call Phil and Betty love each other, very very much. They have been performing some anatomical acts that make it likely that sperm cells from one of them will make contact with an ovum cell of the other. One sperm cell fertilizes the ovum, so it's now called a zygote. Once this happens, you come along and grab that zygote.

Here's the technological part: you nip out the nucleus of that zygote and throw it away; then you nip out the nucleus of one of those cells from the culture plates, which contains the same 46 chromosomes I've been using all my life in every one of mine. Pop its (or rather my) nucleus into that momentarily empty zygote, which can be coaxed to continue operating as such. (Hilariously, one of the techniques at this point is to zap it with electricity.) The point is that the zygote, even without its nucleus, isn't actually empty; it's full of proteins and structures that know how to act like a zygote in ways that my little original cell could not have done (Figure 3.2).

We're using SCINT to get a new zygote using the genes of an adult animal, because the hypothetical "me" in this example apparently wants a twin brother that he didn't get in the initial go-round.

Now, the music swells majestically or sinister, as you please: this zygote undergoes a few cell divisions, just as it would in Betty's reproductive tract, and it is then returned there, or, if she prefers, to some other woman's, although we will keep it

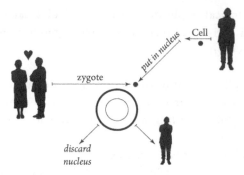

Figure 3.2 Cloning using somatic cell nuclear transfer (SCINT).

simple and say that she is still involved. If you can get it to implant into her uterine lining just as if it had not made a trip to the laboratory, it will do what a zygote does in its accustomed place: develop into an embryo.

In due course Betty's pregnancy comes to term and she delivers a baby. This is my clone. He shares 100% nuclear genetic identity with me. Biologically speaking, this is arguably one of the most boring things we could have spent money on—humans *already* occasionally produce identical twins. There is nothing in any of the biology involved which isn't something that cells, organ systems, and individuals aren't doing anyway. The difference is simply and only a matter of timing: without this technology, genetically identical twins are born from shared pregnancies, and therefore are sociologically siblings as well as biologically—whereas in this case, we aren't.[9]

That is precisely what triggers the alarm: the baffling social factor. Is he Phil and Betty's son? Is he mine? Or is he my brother, and if so, are my parents his? Who is his next of kin? Are Betty, Phil, and I now to be considered legally related? Do we share obligations toward the child? If so, should we come to some parting of ways, who has custody? I like to ask these questions directly in class and watch the students' faces. They always reveal the same thing: at first the turning mental wheels, and then the dawning recognition that they don't know. The reactive crisis arises specifically from the fact that no one knows how to classify such an individual socially.

What is this person's relationship to me? To Phil and Betty? Does he even have a family? Is he legally a person? I'm saying that I, my students, you, and everybody else on this planet, at this instant, have no idea. We simply don't have the vocabulary, the precedents, the cultural history, or to employ a vague but useful term, the value system to answer these questions. I do think they are important questions, due in part to extreme issues like whether I can harvest this person's organs to replace mine at will, but especially mundane next-of-kin issues such as inheritance or custody.

When you can't classify direct evidence into existing social and economic categories, your mind goes into self-defense mode, constructing fabulous connections to justify the contradictory content or walling parts of itself away to process different details in more comfortable isolation. If these constructs are challenged, then your mind typically goes into what Leon Festinger called "cognitive dissonance," an uncomfortable state that may feel like a hostile attack upon you, from somewhere or by something. You'll demonize a certain set of the information, giving it properties it doesn't have but which carry a lot of emotion, and you defend this imagery with some heat. Under further pressure, you'll feel threatened, fearful, and angry, and you'll interpret the surrounding reality as filled with enemies in league with the demons or with insanely stupid people.

Fortunately, due to long years of refinement of my class discussions and exercises, these latter effects don't show up among my students very often, but I do ask them about it, and they do acknowledge the instant potential for the conversation to go right off the rails.

And that is where the "don't meddle" story comes from: because people have encountered products of science specifically as a shock to their value system and social norms, and can slot the provocative input right into the mental bucket called "sorcery." By the time anyone has even begun to speak, they are already in the grip of cognitive dissonance and its associated defense mechanisms, and, for example, see no contradiction between technology they are used to or benefit from, and this new thing, which must be obviously and villainously different.

Anything in the sorcery bucket doesn't merely hold the *potential* for accidents or wrongful use, it simply *is* wrong, shot through and through with abomination. And as such, everything about it may now be tagged as wrong, either obviously so or, despite a benign appearance, lurking in wait. From there, perceiving sociopathic motives, lone-madman characterizations, large-scale conspiracies, or imminent moral and social catastrophe is an easy step. Logic has nothing to do with it; logic is long out the door once this reaction and articulation are established.[10]

In the films, the Prendick character is always altered from his identity as a fellow scientist who differs in subcultural and political history, usually to a practical, everyman identity, with more extreme political and national coding.

In *Island of Lost Souls* (1932), Edward Parker (Richard Arlen) is a hard-nosed, very American engineer, in contrast with Moreau's Australian, colonial character.

Terror Is a Man (1959) features a similar character, William Fitzgerald (Richard Derr), contrasted with the French Dr. Girard.

The Twilight People (1972) features an extra-tough, American warrior type in Matt Farrell (John Ashley), and the ambiguous, possibly British Dr. Gordon.

In *The Island of Doctor Moreau* (1977), both characters are British, with a more sensitive but also tough Andrew Braddock (Michael York).

The Island of Doctor Moreau (1996) features Edward Douglas (David Thewlis), who is a generally ineffectual British diplomat with implications of idealism (he was involved in "the peace talks"), whereas Moreau and Montgomery are American.

Doctor Moreau's House of Pain (2004) features Eric Carson (John Patrick Jordan), an American boxer, against the very British Moreau.

In all cases except possibly the 1959 film, the hero unequivocally and correctly judges Moreau to be despicable. Standout moments include Parker punching him and Douglas's line, "Has it occurred to you that you've completely lost your mind?"

Now for the real example: Dolly the sheep, born in 1996 due to the research efforts of Ian Wilmut and Keith Campbell, who has endured more textual attention than any other sheep in human history and serves as the veritable poster child for cloning. Everything I said about Phil, Betty, me, the possible surrogate-womb woman, and the kid applies here, trading out the humans for sheep. I won't even bother to re-present the diagram. Do remember that part about using a cell from an adult sheep as the donor for the new zygote, though; it's important.

Let me reveal one thing I don't see much in the oceans of references and blog posts: that Dolly was well worth the scientific publication space, but not biologically incredibly interesting. Or rather, that the technology was impressive, concerning a specific logistic point: that the nuclear donor cell had come from a relatively differentiated cell rather than a stem cell. This isn't trivial—it means that you can do what I described earlier for myself, get a twin for a sheep who is already grown up. But it's—you know, just a twin.

Dolly wasn't even the first technologically cloned animal; if you count early embryo splitting technique, then that honor goes to a nameless newt back in 1901! If you want more intracellular methods, then look to a similarly nameless carp in the early 1960s, and for SCINT mammals, to Masha the mouse in 1986, whose legacy might be understated in certain venues because she was a communist. Nor was Dolly even the first SCINT cloned sheep; the same lab cloned two others named Megan and Morag using embryonic donor cells, understandably, considering that you want to get the technique right first before messing with it further. Dolly was the first SCINT cloned sheep whose donor cell was from an adult animal, and biologically, that's *all*. The project that produced her successfully overcame a logistic difficulty, and technologically, *that's* all.[11]

And yet, what outcries were raised! Gina Kolata's breaking-news article in the *New York Times* is a perfect example of factual statements in sensational context, which I suppose was her job, after all. To their credit, Drs. Wilmut and Campbell provided accurate and levelheaded quotes for all their media statements. Kolata also quoted a single line from Dr. Lee Silver: "It's unbelievable. It basically means that there are no limits. It means all of science fiction is true." The full interview shows that Silver was contrasting scientific and popular narratives and was not actually proclaiming this statement, but the single line made headlines.

The coverage and cultural identity of Dolly then underwent a meaningful shift. Since she inconveniently looked like a run-of-the-mill Finn-Dorset sheep and therefore failed badly to play her part as an abomination, other images juxtaposing humans and industrial labware swiftly crowded her out to illustrate the required narrative. The transition was explicit, from *Time* magazine's close-up of two sheep's' faces captioned "Will there ever be another you?" to *Newsweek*'s charming image of happy human babies in scientific beakers, captioned "Can we clone humans?" You can look up whatever you'd like from there; the prose swiftly leaves the range of purple and goes into ultraviolet.

Once the sorcery button was pushed, we the public were treated to way too many confirmations that the academic degree process isn't perfect, like the physicist Richard Seed promising to clone humans and thus make them, or us, for reasons that remain obscure to me, closer to God; the chemist Brigitte Boisselier and her Raëlist associates, who claimed to have already cloned a human and named her Eve (they hadn't); and veterinarian and genetics researcher Hwang Woo-Suk, who published fraudulent claims to have cloned human stem cells in the prestigious journal *Science*. You want to know the funny part? The Dolly technology *didn't* bring us "closer to human cloning," because Masha already did the heavy lifting to make SCINT work, and mice are more closely related to humans, that is, primates, than sheep are. That means that every one of these claims, some or all of which helped to jack the stem cell issue into the political red zone, was hot air.

At the very least, you'd think the notable absence of hordes of SCINT-cloned sheep using adult donor cells, and the confirmation of low success rates for other species since then, would imply that even with the technology, we aren't exactly ready for the champagne yet.[12]

Back to Reality

Biotechnology is uniquely vulnerable to being coded as unnatural when it challenges existing values, and it is similarly difficult to drag out of the sorcery bucket once it's there. I think that's why the novel *The Island of Doctor Moreau* has been so little processed, both culturally and academically. It is indisputably about biotechnology, but it's important to separate that word again. I am arguing that it is not a fear-based novel of the unnatural violation of biology (i.e., don't meddle), but a reflection-provoking novel administering the technological shock to exist-ing values. The difference is profound: in the first, the proposed science *can't* be done, because it's unnatural, and you try to stop it, or if you can't, to kill it or escape it, and regardless, it will all collapse into ruin without fail. Whereas in the second, it most certainly *can* be done, in accord with existing natural principles, and you are forced to confront your existing values with the evidence that they are not as established as you thought.

I suggest that *The Island of Doctor Moreau* is the second type, and it deserves a fair reading, with the "don't meddle" model set aside. The characters are vivid if you let their voices speak in your mind, and their decisions matter, changing the situation rather than playing through an obvious and familiar plot. Moreau's work does not go "against Nature," but rather manipulates it well enough to accom-plish what he does—the question is whether he can see it, which he cannot, and whether Prendick can, which he does all too well as far as he's concerned.

Unlike thrillers, science fiction and otherwise, which ultimately confirm the audience's values and sense of equilibrium, this particular kind of science fiction uniquely does the opposite: to expose a taboo, to disturb one's equilibrium and even identity. It will repay you beyond any novel I have ever read or know of.

Readings

Abigail Burnham Bloom, *The Literary Monster on Film* (2010), analyzes transitions from literature to cinema for five Victorian novels, generally concluding that the nuanced and personal conflicts in the former are converted to more general threats and fearful imagery in the latter.

A long-running analysis of science on film begins with George Gerbner, Larry Gross, Michael Morgan, and Nancy Signorielli, in "Scientists on the TV Screen" (*Society* magazine, 1982), suggesting that its presentation negatively influences exactly those people who are otherwise most favorably inclined toward science; see also their more complete article "Science on Television," *Issues in Science and Technology* (1987), which includes the statistic that television scientists get killed far more frequently than any other profession or designated social role. Other use-ful analyses are found in Sidney Perkowitz, *Hollywood Science* (2010); David A. Kirby, *Lab Coats in Hollywood* (2013); and Christopher Frayling, *Mad Bad and Dangerous? The Scientist and the Cinema* (2005).

Henry Petroski, *The Evolution of Useful Things* (1994) and *To Engineer Is Human* (1992), provides a crucial, non-idealized perspective on human invention and technology. I can't say that Alvin Toffler's *Future Shock* (1970) is all that deep, but it is a pop culture classic and worth reading to see where the meme came from. See Elaine Tyler May, *America and the Pill* (2011), and Jonathan Eig, *The Birth of the*

Pill (2014), for readable introductions to the history of this invention. The concept of cognitive dissonance was introduced by Leon Festinger, *When Prophecy Fails* (1956), and is reviewed in Joel Cooper, *Cognitive Dissonance: 50 Years of a Classic Theory* (2007). I find that these three bodies of work produce a powerful effect in combination, more so than any of them alone.

The best discussion of Dolly comes straight from the scientists, Roger Highfield and Ian Wilmut, *After Dolly* (2006). Of others' popular accounts, the better-researched include Gina Kolata, *Clone: The Road to Dolly, and the Path Ahead* (1997); Sarah Franklin, *Dolly Mixtures* (2007); and Stephen Levick, *Clone Being* (2003). Plenty of people have provided thoughtful reflections on cloning and SCINT cloning, but we have little or no societal mechanism to bring such reflection forward and into the decision-making sphere, as opposed to more electorally valuable, decision-freezing hype.

Dr. Lee Silver has written extensively on the interplay of technology and values; his *Remaking Eden* (2007) outlines the shock to values presented by biotechnology in more detail than I address here, and his *Challenging Nature* (2007) demonstrates that the aversive reaction to such technology can be found across the full range of the American political spectrum.

Notes

1. In her discussion of *Frankenstein* (*Frankenstein and Radical Science*, 1993), Marilyn Butler explains why Victor Frankenstein does in fact feature a few "mad scientist" tropes, similarly emerging from its effectively invisible practice just as it does today.

2. The "don't meddle" story is less common in science fiction literature, but became more so during the 1980s when the genre became politically more mainstream and book publishing became tied more tightly to film production. Special mention goes to Michael Crichton, who specialized in presenting the "don't meddle" story over and over for whatever fear of science *de joeur* reigned at the moment. Crichton had the opportunity to defend his views at length to the American Association for the Advancement of Science in 1999, which is available online at http://www.abc.net.au/science/slab/crichton/story.htm. I leave the assessment of his argument up to you. Not all stories that criticize science or depict it in scary ways are of the "don't meddle" type. It makes perfect sense to use technological imagery to exaggerate current issues into fantastic shapes, the better to bring those issues into the foreground, especially when they are taboo or too-well embedded in marketing or political power to be addressed literally. Politically dissenting stories, or psychological dramas, or anything like that put into science fiction terms are fine and fascinating things. But such stories are not "don't meddle" policy statements. The horror in *GATTACA* isn't biotechnology, it's discrimination. I even admit that some "don't meddle" stories are pretty good, especially when a bit excessive. Great examples include *Shivers, Re-Animator,* and *Altered States.*

3. In the 1980s, Gerbner suggested that the depiction of scientists is gradually improving and that the negative image is being rehabilitated, but I don't see it. What I see is that the positive scientists are of a specific, task-oriented type who know their place and never meddle.

4. I like to ask my students which of the two primary characters in *The X-Files* is the scientist. Most answer "Scully," based on the show's own terminology or claim, which provides the

opening for my observation that she responds to mysterious events with a marked aversion for observation and analysis, preferring to refer to texts as bodies of established and completed facts. In this issue at least, the show therefore manages to provide a sympathetic view of active scientific thinking by coding Mulder as the non-scientist in cultural terms.

5. Philosophically inclined readers will recognize this notion of intrinsic blasphemy in the family of concepts collectively called the "naturalistic fallacy," especially in the virtual identity posed between the concepts of nature and God.

6. A little sting lies in the tail of the novel's history. At the end of *The Island of Doctor Moreau*, Prendick turns to two scientific topics he considers sufficiently pure and abstract to be of no danger to anyone: chemistry and astronomy—the very two that would become atomic theory and relativity, and therefore within a few decades, would yield nigh inconceivable technological atrocity.

7. I suggest that the sexual revolution threw every social group in the United States and elsewhere for a loop in the 1960s, specifically including the counterculture and feminism.

8. UN Convention on Biodiversity, published by the United Nations, 1992 (available at https://www.cbd.int/doc/legal/cbd-en.pdf).

9. A SCINT clone isn't even a particularly "identical" identical twin, because the only identity between the two individuals lies in the nuclei. Non-technological identical twins share cellular identity, from their initial zygote's first cellular division, including all sorts of proteins and maternal mitochondria in the cell, and at the organismal level, the profoundly important environment of the pregnancy. This little baby doesn't share any of that with me. He is likely to look like me in a number of variables, but to expect him to develop into an anatomical and physiological *copy* of me or to call him "Ron Two" would be impressively stupid.

10. A single episode of the TV show *Farscape*, "DNA Mad Scientist," is "don't meddle" to a remarkable degree, and bears watching as a near-perfect example. The background of its production merits investigation, as this one episode's content diverges sharply from the otherwise enthusiastic characterization of scientific thinking throughout the rest of the series.

11. Cloning and genetic engineering are two different things. Neither Dolly nor my hypothetical human example includes genetic engineering.

12. I also uncharitably point to the 276 failed attempts accompanying Dolly's successful implantation and birth and the 242 failed attempts that accompanied Megan and Morag. All SCINT cloning is plagued by lousy returns, for reasons at several levels. The nuclear transfer is severely intrusive to the cell, such that only a fraction undergoing the procedure survive; in those that do, apparently the genes themselves suffer damage to varying extents; and of the very few animals that develop from the cloned cells, even the lesser levels of genetic damage they've incurred typically induce a number of physiological disorders. All of this was known already, well established by the work on mice.

PART II

THE THING IS AN ABOMINATION

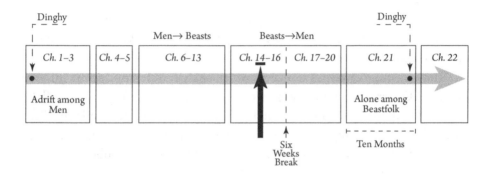

Men are the devils of the earth, and animals are the tormented souls.
Arthur Schopenhauer, "On Religion," *Parerga and Paralipomena*,[1] p. 187

When I was eleven, my mother and I regularly visited local nursing homes, she with her sheaf of music to play on the cheap electric piano and I with my trumpet. See us there: we play for the residents, all of them helpless with different conditions, most attentive to us, some unable to speak, few with the ability to let us know what they hear or think. Some try hard, and we listen, nodding, holding their hands.

Ten years later, I hold a living rat in my hands. He knows me and sits comfortably. He had undergone surgery weeks ago, under full anesthetic, his sutures treated with analgesic until they healed. Now, there is a metal implant in the medial preoptic area of his hypothalamus and another in his abdomen. I had checked his activity, food and water intake, and responsiveness every day. If at any time he had been listless, moved spasmodically, stared fixedly, or been off his food, I would have killed him. I watched him mate, recording his and his partner's behavior and vocalizations, a radio pickup recording the two temperatures broadcast by the implanted equipment. Now, I put him in a plexiglass container and pump in carbon dioxide—he dies instantly, without kicking or twitching, his bladder remaining full. I make absolutely sure by cutting off his head.

Around that time, my mother lets me know that she wants to be euthanized if she were to become incapable of self-maintenance. The laws in the United States are stricter than she likes, but her arrangements have pushed them as far as she can. She tells me that she can't quite trust my brothers to go through with it, so she's signing full power to me as the executive in this matter. "You understand," she says. "You will be able to kill me." If that day comes, I will hold her hand.

The question of ethics concerns what shall I, a person, do in this life I have. Questions of policy, however, means negotiating what will be applied and enforced, and in which communities, and how reliably. They are not the same thing. No one wants policy with no ethics in it, but policy isn't made simply by agreeing on the ethics. One might say policy gets made because we *can't*.

What policy governs scientific research? It's animal use to be sure, but a different use from burgers or from enlisting animals in labor. It's intellectual, assessed by people talking and thinking, rather than by eating or working. It looks alarming in its instruments and sounds disturbing, with words like "control" and "subjects." The clinical observation, sharp categorizations, and the literal intervention into functional tissue add up to a real "hot button." It's easy to startle backward, saying, "This is wrong!" It's harder to ask what responsibility we will assume over death and pain, without fear.

Note

1. Arthur Schopenhauer, "On Religion," in *Parerga and Paralipomena*, 1851, published as *Essays and Aphorisms*, edited by R. J. Hollingdale, Penguin Books, 1970.

4

The House of Pain

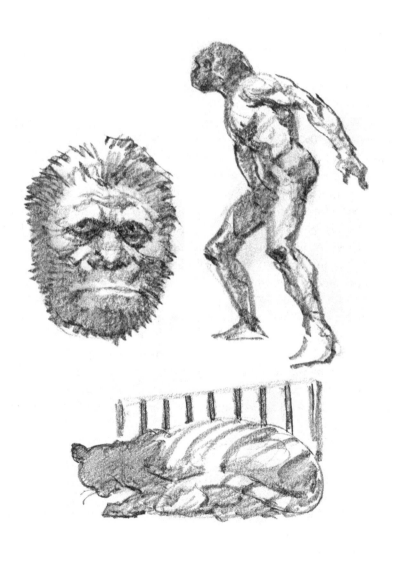

Moreau straps a living animal to a table. He cuts its body and manipulates the tissues into new arrangements. He immobilizes the animal's head and ties its jaws open, then uses long needles to penetrate into its brain through the top of its pharynx, the shared part of the nasal and oral cavity. The animal is permitted to heal to some extent, then he conducts more interventions for further manipulations, until its healing results in a form much like our own. The animal fully experiences the pain every time. It can do nothing but scream.

Moreau's project is an abomination, yes. What does that mean? What if he had used anesthetics, and used them well, so the procedure was free of pain? Just as wrong? Less wrong? Not wrong at all?

The agony is an issue all its own. Prendick never approves of the pain. After Moreau is killed, and after he and Montgomery recover his body, the first thing they do is to kill every subject currently in the surgery. He goes through a lot of changes and shifts of attitude about the humanity of the project's created people, but not about what they suffered during their transformations, which always disgusts him. Yet he cannot say why, nor can he refute Moreau in their debate in Chapter 14. Significantly, until Moreau dies, he does nothing to restrict or stop the surgeries, tacitly accepting the project.

Why? Does Prendick have a solid moral argument against Moreau? He's not a helpless captive, and Moreau has no research intentions toward him. During their conversation, Prendick is armed, and their entire debate is embedded in the freedom that Moreau has granted literally to *shoot* him if he cannot justify his position. I know plenty of people who would unhesitatingly put two bullets in Moreau's chest and one in his head at the end of that conversation. Why Prendick doesn't do it requires some reflection, both on the history at the time of writing and on the issues that persist today.

> Most of the films drastically alter this dynamic, making the hero explicitly into Moreau's prisoner, in turn making their debate or discussion unequal, as the scientist could always respond vindictively to being refuted. Typically it's expressed by the hero discovering that his room is barred and locked in a far less ambiguous fashion than the way Prendick's room is secured in the novel, and in some cases, literally penning him into a cell when the scientist has experimental designs upon him.

No Pain, No Gain

Until the 1790s, the word on medical pain relief is simple: nothing existed worth the name. If you were to receive surgery, you simply—well, you got to participate. Surgery on humans was in transition from surface features and wound repair to opening and closing the abdomen. That transition lasted decades, roughly from the first consistently successful ovariotomies in the early 1800s until the first excision of the gall bladder and the first abdominal appendectomy in the 1880s. I find it impossible to contemplate the horror of the experience, and I shudder at the recovery rate and the recoverers' morbidity rate, given that sterile technique, although suggested as early as the 1790s, wasn't widely adopted until a century later. Consider that the American physician James Marion Sims developed the surgical correction of vesico-vaginal fistula in the 1840s and 1850s, but his first success required *thirty-three* anesthesia-free operations on the *same* three patients.[1] Bear in mind as well that shock was not studied until the 1890s, and often was disastrously treated with stimulants, if at all.

That makes three sets of techniques in uneven development: shock, sterile technique (without antibiotics, this is confined to assiduous washing-up and wound cleaning), and anesthesia. These were also debated and developed in the context of population and policy problems, including the toll of rapidly developing technological imperial warfare on soldiers, the terrifying incidence of puerperal fever in maternity wards, and epidemics of typhus, cholera, and typhoid fever in London alone, as well as the epidemics among displaced and crowded people throughout the colonies.

The most relevant feature to Moreau's work is pain alleviation, including general anesthetics. You can forget all that business about berries and herbs, and about alcohol, which is best understood as a sedative—it doesn't dull pain so much as make you less able to struggle free while people saw off your leg, and the necessary quantities are as likely to produce uncontrolled and dangerous vomiting as quiescence. Other powerful inhibitors such as hashish, mandrake, and opium bring a high chance for fatal overdose in the necessary quantities for surgery.

In the second half of the nineteenth century, inhaled anesthetics were at least available. Both nitrous oxide and ether (technically, diethyl ether) had been isolated earlier, but seem to have been used mainly as party drugs. Their use as anesthetics didn't begin until the 1840s, and nitrous oxide turned out to be effective mainly for dental work. The first reliable general anesthetics were ether and chloroform, which had been discovered in the 1830s.

At first, a patient's chance of dying may have increased rather than decreased. The problem is simply that all general anesthetics are by definition poison: having so many nerves inhibited is flatly not good for you, in more ways than I have room to list, and—like alcohol—both of these are crudely powerful and individually unpredictable. The dose of chloroform required to keep a patient anesthetized is very close to the lethal level. Ether is a little bit safer in terms of overdose, but it's more damaging to the patient's lungs; furthermore, its wearing off entails such nausea and so much vomiting that dehydration or disruption of one's sutures—that is, massive internal hemorrhage—are a mortal threat.

Until technology could provide a more precise way of monitoring the dose throughout the procedure, and also to monitor the patient's vital signs, any operation using general anesthesia was a race between the surgical techniques and the patient's death due to the anesthesia or shock, and later, to postoperative hemorrhage. Controlled inhalation techniques combining nitrous oxide and ether were developed in 1876, but even then, anesthetized surgery was automatically life-threatening.

For people who could afford constant attendance, and especially for people whom the physicians would be terrified to lose, nitrous oxide and ether might be relatively safely available—emphasis on "relatively." Queen Victoria famously delivered Prince Leopold under ether in 1853, quite early in this history, affording alternate-history buffs the opportunity to speculate about what might have happened with the British Empire if her dosage or response had been a bit off. But for low-income patients being treated rapidly and in great numbers, without the necessary technology, training, and infrastructure for supply and expense, anesthesia was both less available and much more dangerous.

Our notion that both pain relief and safety should be maximized developed much more recently than I find comfortable to contemplate. It wasn't until Henry K. Beecher published *Physiology of Anesthesia* in 1938 that modern substances and techniques began to be used, and I still recall from my childhood in the 1960s and 1970s that people who underwent major surgery either died or returned dangerously underweight and physiologically ten years older.

This technological history provides the context for pain relief in nonhuman animal subjects and patients as well. Consider William Harvey's famous experiments on circulation way back in the 1620s, which accurately identified the double circuit in the movement of mammalian blood, heart-body-heart-lungs, currently taught in health and biology classes everywhere, all the time. You may have encountered it in *Schoolhouse Rock* or *The Magic School Bus*. But modern education has more or less forgotten that this work relied upon horrific agony for the many animals he used, and the list of historical experiments underlying many such basic points of health and biology includes countless examples. Anesthetics simply didn't exist.

The 1850s and 1860s were the transitional period, when the medical practice was becoming a profession with a much wider audience and more precise curriculum, and experimentation and general knowledge of normal function were immediate policy concerns. General anesthesia was possible if variable and unreliable, investigation included lab work on healthy bodies as well as clinical observations, and the culture of research had neither a moral nor legal precedent for considering the subjects' experience to be important. Most, although not all, researchers using live animal subjects considered anesthesia little more than a likely way to bias the results or to lose the subject and hence the data halfway through. That's where the physiological researchers were coming from, in stating so strongly that seeking scientific knowledge automatically must accept inflicting pain on experimental subjects, or "no pain, no gain." It's not *true on principle*, but it was the historical *case*, or *logistic situation*.

The case study must certainly be the French physiologist Claude Bernard, both in general and specifically as presented in the chapter addressing experimental subjects' pain in his book *An Introduction to the Study of Experimental Medicine*,[2] published in 1865:

> The science of life is a superb and dazzlingly lighted hall which may be reached only by passing through a long and ghastly kitchen. (p. 15)

[and]

> A physiologist is not a man of fashion, he is a man of science, absorbed by the scientific idea which he pursues: he no longer hears the cry of animals, he no longer sees the blood that flows, he sees only his idea and perceives only organisms concealing problems which he intends to solve. (p. 103)

If Réne Descartes' statement that animals were soulless, clockwork mechanisms incapable of pain was ever taken seriously by scientists, which I doubt, it wasn't commonly invoked during the heyday of vivisection. Bernard and contemporary researchers did not claim that animals feel no pain, and should at least be understood as not hiding behind denial.

Here's an example. During this period, glucose (blood sugar to you and me) could be reliably assessed by a technique called copper reduction. By 1848, physiologists had recorded the primary insight of digestive physiology: when you eat carbohydrates, glucose is soon found throughout your circulatory system; it's pretty clear that we get it from what we ate. The question people wanted to address was, is that it? More technically, do we synthesize glucose in addition to consuming it?

Actually I'll cheat and give you the answer first (Figure 4.1). Your stomach and intestines have carbohydrates in them, because you ate food. They get digested into the littlest carbohydrates in there (that's what digestion means, hammering big chemicals into little ones), small enough to be absorbed into your blood, mostly as fructose and glucose. All these blood vessels at the intestines are collected into the hepatic portal vein, delivering the blood to the liver. Then the hepatic vein delivers it from the liver to the general circulation of the body, specifically the vena cava, just before it gets to the heart. Or to put it differently, blood at the heart has only the amount of glucose in it that the liver permits. The liver is the gatekeeper for the glucose, because instead of just passing it on with the blood, it grabs it and stores it in a form called glycogen; the blood that goes into the hepatic vein gets glucose that's resynthesized from the glycogen. Everything you've heard about "blood sugar" is a function of what the liver is doing with its glycogen/glucose conversions.

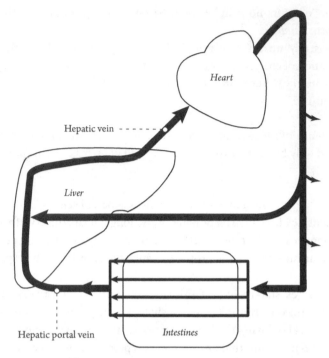

Figure 4.1 Hepatic portal system.

To figure this out, a whole lot of preexisting assumptions had to get over-turned. People already knew that the liver produces bile (a digestive facilitator for fats), and, incorrectly, "one organ one function" was the prevailing dogma. More subtle is the idea that the sugar did have to get consumed, as we cannot make it *de novo* from simpler compounds (like plants can), but we store it in a fashion that means synthesized sugar is what we actually use in our tissues. Conceiving of body systems so that such ideas can even be addressed, and then taking the questions of how such things are chemically regulated, represented a whole new universe of thought. The credit for spotting these assumptions and setting up a more functional model for investigating blood sugar acquisition and regulation belongs to Bernard, from work reported in his *Leçons sur le diabète et la glycogenèse animale* (1877), published by Cours du Collège de France.

Bernard does not report what animal he initially used, but I'm assuming it was a rabbit; the laboratory rat and mouse had not yet found their place in science. The first step is to withhold food from the animal. Then he restrained it, opened the abdomen, pushed the intestines aside, partly folded up the liver, and withdrew blood from the vessels going into the liver from the intestines (hepatic portal vein) and also from the vessel (hepatic vein) going from the liver into the vena cava (the big vein going back to the heart). There's glucose in both places.

More animals are needed for the first investigation of cause. This time, he tied off the hepatic portal vein to find that glucose is still present in the hepatic vein; in other words, the liver is producing glucose in some way. For this to be assessed, the animal stays that way, its abdomen opened, for hours. Bernard repeated this across a broad comparative array of vertebrates, which I presume means cats, dogs, and other livestock, and found the same throughout: liver tissue full of glucose, and nowhere else except as supplied by the liver. Please imagine a veritable parade through the laboratory of many species of strapped-down, muzzled animals under no anesthetic, with their abdomens open, their intestines reflected, and the liver inverted for access to the blood vessels.

So the liver is more than a mere speed bump between the intestines and the vena cava; it's somehow delivering glucose on its own. But we haven't yet answered the question of whether the liver stores the glucose it receives from the intestines, or makes it from scratch independently from digestion.

An ordinary if unpredictable detail of scientific work is the semi-accidental discovery. In this case, Bernard examined the contents of a liver that had been left sitting out all night, and he discovered more glucose present than its previous result. This suggested that the liver might be making the stuff, so his next experiment was to open the animal and inject water into the portal vein entering the liver, until the blood leaving it via the hepatic vein is "washed out" (i.e., has no glucose in it), and then assess that same vein to see if glucose reappears. He indeed found that this happens, strongly supporting the surprising notion that the liver is indeed making glucose, and coined the name "glycogen" for the precursor substance. Note here again that the animal is lying restrained, alive, gutted, and in pain even longer this time—at an informed guess, at least twelve hours, to permit multiple timed blood draws from the two veins.

This conclusion revolutionized the study of the body. First, it demolished the long-standing doctrine that a given organ has a single function—here, the liver not only produces bile, it makes and delivers glucose to the primary circulation of the body. Second, it changed how questions were asked about organs, in that they not only "do" something, they trade off and vary the rates among many functions, in a reactive and interactive way, connected by a system of signals.

Bernard's next question is precisely the right one, based on this new perspective on bodies: What prompts the liver's production of glucose into the hepatic vein? This is subtler than it looks, because instead of assuming that organs do things because they "know" or "should," he's looking for a systemic interaction among body parts—signaling based on current conditions, what today we call regulation. It's also striking that he turned to the nervous system as the probable regulatory agent. He focused on the vagus nerve, as it was already understood to have powerful inhibitory effects (it regulates your heart rate this way, for instance), and he cut its branches that go to the liver, finding that this did indeed result in less glucose being produced. However, stimulating the nerve in this location did not increase it, failing to provide the positive control that would lead to a strong conclusion.

To address the nerve effect more directly required a second technique besides opening the abdomen: to go into the brain through the soft palate. You know what the soft palate is, right? Put your fingertip on the roof of your mouth behind the front teeth, then slide it backward toward your throat—when you hit the squishy part, that's it. Bernard incised the soft palate and used a needle to stimulate the underside of the fourth brain ventricle, where the vagus nerve begins. I trust you are imagining a fully awake animal restrained with its jaws at maximum gape, such that a person can reach its brain tissue through its soft palate, in addition to the abdominal intervention. He found that stimulating the vagus makes blood glucose increase dramatically but briefly, and that a more radical intervention, cutting the spinal cord to block the (stimulatory) splanchnic nerve, does the opposite, leading him to suggest that the opposed effects of these nerves regulated the liver's release of glucose to the body.

This turns out to be incorrect, as interfering with these nerves also brings adrenaline into the picture, one of several hormones prompting an increase in blood sugar from the liver. That's why his results included the confusing effect from either stimulating or cutting the vagus's terminal branches at the liver, as well as the similar detail that cutting the vagus at its origin had the same effect as stimulating it. Specific and immediate conclusions are not the currency of science, though. Since the results included both valid conclusions and an ambiguous effect, new debates ensued that opened up the whole idea of hormonal regulation occurring simultaneously with neural regulation. Bernard's work set up an understanding of whole-body regulation to an extent never before imagined in the study of physiology.

If you're looking for moral judgments from me, I'll give you one. The person who conducted that protocol was a monster. Not only do I wish he had been prevented from doing it in the first place, but that if he did it, he'd be punished. In fact, I would like to dig him up and gibbet his body at—let's see, what would be my local equivalent of London's Blackwall Point—I suppose out at that construction site by the interchange that never seems to get finished.

A lot of people back then thought similarly, as Bernard was well known to activists and was regarded as little better than Satan himself. Although he lived and worked in France, his work was widely followed and so influential on physiology and medicine that the Royal College on Gower Street could well be considered a branch of his lab, and his blunt acknowledgment of animal use made his name the embodiment of vivisection in activist terms. You can also see him in the physician Edward Berdoe's frightening and popular novel, *St. Bernard's: The Romance of a Medical Student* (1887), published under the pen name "Aesculapius Scalpel," followed by a second volume with annotations explaining the sources for the fictional events as *Dying Scientifically: The Key* to *St. Bernard's* (1888). There's no way the fictional horror hospital's name is a coincidence.

Bernard's outlook wasn't unanimous among physiologists, and according to *Applied Ethics in Animal Research* (editors John Gluck, Tony Pasquale, and F. Barbara Orleans), he did eventually adopt anesthetics for his experiments, but scientific advocacy for anesthetics tended to get muddled or drowned out. Marshall Hall was in some ways his English counterpart and was criticized as a vivisectionist, but he did propose self-regulatory principles, which acknowledged that the subjects were sentient (his term). The head of the Royal College at the time, J. B. Sanderson, spoke in favor of regulation, but his 1873 *Handbook for the Study of Physiology* did not include anesthesia and became notorious. During the hearings of the Royal Commission, George Hoggan, an English physician, criticized cruelty and waste in Bernard's laboratory in a letter to the *Morning Post*.

Most of the physiologists' testimony heard by the commission invoked his successes and practices as the height of scientific achievement, and not dishonestly. If Bernard's work was analytically unsound, he might be written off as a grotesque historical curiosity. However, it was not only sound, but foundational. *An Introduction to the Study of Experimental Medicine* set the intellectual bar for scientific technique and reasoning as high as it has ever been relative to contemporary practice. He championed and demonstrated the scientific method, including such important concepts as the relationship of conclusions to certainty, and the double-blind experiment. In addition to the liver processing of blood glucose to glycogen, his results include the discovery of homeostasis (he called it the *milieu intérieur*), pancreatic digestive enzymes, vaso-motor nerves, the beginnings of understanding the endocrine system, the coordination of processes across multiple organs, and more. He may have been personally responsible for the most effective quantum leap in knowledge of vertebrate body functions in history; probably every chapter in a standard textbook of undergraduate training in physiology can trace its contents to one or more of his experiments.[3]

I have taught literally thousands of students either straight-up physiology or anatomy with multiple physiological tie-ins, and I've taught just as many about hypothesis testing and experimental reasoning in exquisite detail. Without knowing it, I quoted Bernard up, down, left, right, inside, and outside.

Why didn't I know? For some topics, biology texts border on the hagiographic when picturing and describing the foundational work of historical scientists. In teaching physiology and statistics, I never saw one picture or one mention of Bernard by name. Granted, the physiology texts tend not to showcase the historical personalities and experiments much anyway; it's not as if there were tons of other scientists that were mentioned. But statistics and core biology texts often do, and Bernard's explanations of hypothesis testing may be found practically word for word in both. He isn't in any of those. I can at least speculate that our teaching is too squeamish to admit the details of its experimental history: that although "no pain, no gain" isn't in principle true, it was applied as true during this significant period of discovery.

That squeamishness is not such a good thing: it tacitly supports the value that "the pain was worth it," privileging human gain as an entitlement; and it obscures what researchers do differently today, inviting rock-throwing from those who see only the kid versus the guinea pig in the windows.

Moreau 1, Prendick 0

In Chapter 14 of the novel, Prendick is finally fully informed of Moreau's work, and the two of them settle down for one of the great intellectual cage-matches in literature. Thankfully, it's actually not about sentiment for dogs matched against sentiment for babies. Instead, the two positions are exquisitely precisely chosen, pitting the reform-sympathetic intellectual against the physician, and simultaneously pits the naturalist-scientist against the physician-scientist, regarding the value of their work (Figure 4.2).

Moreau is the fully grounded late-stage medical physiologist, aware of his subculture's continuing victories in the debates in and out of science, and also mining evolutionary theory toward his ends. Overall, he is rather cleanly constructed from several historical sources. Certainly he owes a great deal to Bernard, but there was in fact a French neurologist named Jacques-Joseph Moreau whose ideas factor into the fictional character's techniques. The fictional character is converted to British, despite his name, and his former career is quite rightly and most sensitively located on Gower Street, presumably part of the Royal College.

Prendick is more complex. He is not, as he might have been, a fiery everyman reformer with a street Protestant's intuitive faith, nor a hard-nosed but decent engineer as in several of the films. He is instead a student of Huxley's who references contemporary evolutionary ideas casually and accurately, with a default sympathy toward research, but also overlapping with social reform, especially temperance. His reflections in Chapter 8 place him right on the fence concerning live animal research. Think of him as exactly that educated person who is confronted by Huxley's *Evolution and Ethics*, as it throws his rather sunny view of evolution and human achievement into shadow.

The discussion proceeds through five phases:

1. They bounce back and forth between pain and knowledge, illustrating how the two are historically confounded.
2. They address the goal of achieving humanity, mainly explaining its feasibility.
3. They debate the morality of pain as such, with the subset topic of religion.
4. Moreau describes his early experimentation.
5. They revisit the goal of achieving humanity with an inadvertent self-revealing comment from Moreau.

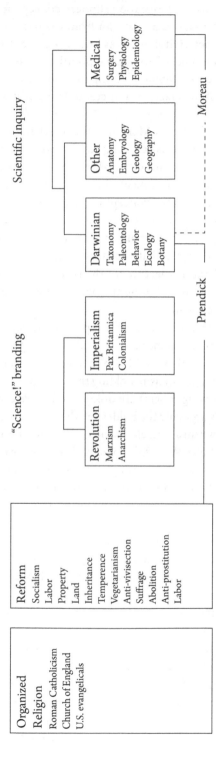

Figure 4.2 Views of science and policy during the nineteenth century.

To them, the discussion hinges on a slippery concept: materialism. By the time of the novel, this word has changed a lot from the days of Abernethy's invective. The topic is less about vivifying forces and more about evolutionary origins— "man coming from monkeys" to use the popular phrase—but as an accusation, the term retains its core content of failing to exalt life and humans, and rejecting morality. Arguably its fear factor was increased by this shift to a less abstract topic than chemistry.

When Prendick objects to Moreau's characterizing him as a materialist, it's because he thinks Moreau is referring to a famous exchange of essays, beginning with the phrase "I am no materialist," in Huxley's *On the Physical Basis of Life* (1868). In the purely technical philosophical terms as used by Lawrence, this essay is as material as the concept can get; however, Huxley is objecting to a new use of terminology and its core ideas, called materialistic philosophy, or the new philosophy, or positivist philosophy, as espoused by the French philosopher Auguste Comte. Huxley hated materialistic philosophy so much that I don't think he presented another lecture for the rest of his life without taking at least a moment to criticize confounding the observations of physical life with moral directives or social obligations.[4]

However, and more relevant here, the topic of materialism had recently received further development in a further exchange of ideas, beginning with Huxley's term "agnosticism" in his 1869 essay, "The Nineteenth Century." His point of view is today called positive agnosticism, holding that metaphysical questions (including "is there a God?" among others) are unknowable, irrelevant, and ultimately not very interesting. It's arguably the boldest of all such views, nothing like the popular notion of agnosticism as merely sitting on the fence between faith and atheism.

The religious scholar W. S. Lilly objected to this idea with some force in his 1886 *Materialism and Morality*, stating a common viewpoint among theologians, that faith in metaphysical principles, both in intention and in their presence in physical reality, was required for a person even to have a moral compass, let alone lay claim to one. He argued that a person who views the natural world, including humans, as composed solely, or even by and large, by physical and understandable processes, is unconstrained from unregulated, sociopathic conduct. In other words, to be ethical at all, one must necessarily embrace everyday reality as full of miraculous qualities and guided events.

In Huxley's rejoinder the next year, in *Science and Morals*, he had not yet arrived at the grimness of his 1893 *Evolution and Ethics*; he replied that the mind is organic, and indeed deterministic in chemical terms, but that science, the product of the mind, can in fact yield morality and ultimately happiness.

Therefore, against Moreau, Prendick is asserting his opposition to Lilly's accusation, nigh perfectly quoting Huxley, although as it turns out, he's misunderstood Moreau's more sophisticated use of the term. It's an important misstep. Huxley was famously successful at puncturing his establishment religious opponents like Lilly, both with logic and by knowing their texts better than they did. I have no doubt that Prendick would have been equally handy in a similar debate. However, he is completely unprepared for matching ideas against Moreau, who mops the floor with him. I'll pull their debate topics into a more schematic order to show how.

Morality and Knowledge

Moreau expertly refutes the twin accusations of insanity and hubris. Regarding his motivation, he rightly claims the mantle of plain ordinary science:

> ... I went on with this research just the way it led me. That is the only way I ever heard of research going. I asked a question, devised some method of getting an answer and got—a fresh question. Was this possible, or was that possible? You cannot imagine what this means to an investigator, what an intellectual passion grows in him. (*The Island of Doctor Moreau*, pp. 55–56)

That's not insane, nor is it intrinsically cruel or unjust. I know every detail of it myself. His question of plasticity even matches my own research topic, and is not at all distant from what we now call Evo Devo (evolutionary developmental biology), or the variations in development that lie at the heart of species variations. This is what he's seen of how the body works, and this is what he seeks to know about it, and as long as we're talking about the "plasticity of the shape," then he's on track. (He veers off that track later in the discussion, though.)

He also scotches the image of the God-defiant, would-be-God scientist effortlessly, and firmly establishes his intellectual bona fides by referencing his grounding in faith.

> ... I am a religious man, Prendick, as every sane man must be. It may be I fancy I have seen more of the ways of this world's Maker than you—for I have sought His laws in *my* way, all my life, while you, I understand, have been collecting butterflies. (Ibid., p. 55)

This might be a little odd for modern audiences, who are strongly influenced by twentieth-century American political rhetoric, so I need to emphasize the metaphysical importance of knowledge to most nineteenth-century researchers. They generally felt that studying Nature's utterly physical mechanisms is effectively worshipping God (the Maker), and are therefore religious in an approximately Deist way straight out of the *Vestiges*. This language may sound a bit strange today, but it was explicitly accepted as a valid scientific outlook in its time, and it might apply to more modern scientists than you think.

Morality and Agony

Granted, Moreau's clarity of thought may not be immediately obvious at the "don't try this at home" moment, when he flicks open a knife and stabs himself in the thigh.

> ". . . Why, even on this earth, even among living things, what pain is there?"
>
> He drew a little penknife as he spoke, from his pocket, opened the smaller blade and moved his chair so that I could see his thigh. Then, choosing the place deliberately, he drove the blade into his leg and withdrew it.
>
> "No doubt you have seen that before. It does not hurt a pin-prick. But what does it show? The capacity for pain is not needed in the muscle, and it is not placed there; it is but little needed in the skin, and only here and there over the thigh is a spot capable of feeling pain." (*The Island of Doctor Moreau*, p. 55)

Is this even possible? It sure is! Well, even the smaller blade of a penknife is probably too big, but yeah, you can do this. It's a standard exercise in undergraduate physiology labs: the students use felt pens to mark off squares in a grid on their forearms, then use pins to poke the individual squares on one another, without the poked person looking, recording which ones register pain and which don't. They're routinely surprised by how much of the surface area doesn't feel it. Pain is a local physiological phenomenon, and receptors for it are not distributed continuously throughout the skin, although Moreau exaggerates the receptors' scarcity, and his act of actually driving the blade into his leg must be tagged as artistic license.

It is possible to misread this as Moreau demonstrating the triviality of pain by ignoring it, but that is not his point at all. He's demonstrating that pain is absent when specific nerves are avoided, to show that pain is not a signal from the heavens that some act or inflicted experience is immoral.[5]

Moreau drills on from there to Bernard's point, where he departs from my (and I imagine many people's) sympathy:

> . . . The thing before you is no longer an animal, a fellow-creature, but a problem. Sympathetic pain—all of I know of it I remember as a thing I used to suffer from years ago. (Ibid., p. 56)

Appalling as this is to me on its face, it is philosophically sound for its specific point. Moreau is saying that inflicting pain is simply and only an individual choice, a matter of whether you will or will not do it, and that its worth is a matter of the goal toward which it's directed. It's not even "the ends justify the means," but rather, "if this is the end, then these must be our means," which is effectively Bernard's position to the letter: if you're going to study creatures, and you say the goal is worth it, then at this time in history, you inflict the pain. In that context, it's morally insoluble—you can't determine the right thing through debate, you simply individually have to choose what you will do, and Bernard chose to seek his Golden Age in his ghastly kitchen. Before you get snooty about that, consider that culturally and historically, we all did it with him.

The Superiority of Applied Research

Perhaps it takes training in ecology and systematics to feel Moreau's other, damning hit fully, but there is no historical buffer in this case. He exerts the rather crude but extremely culturally effective pride of place felt by some medical physiologists, who consider themselves at the apex of scientific thought and societal benefit, in contrast to the waffly "nature special" romance of other biologists. Armed as well with authority over patients and moral certainty regarding health benefits, they find natural history to be trivial. To revisit the earlier quote:

> ... I am a religious man, Prendick, as every sane man must be. It may be
> I fancy I have seen more of the ways of this world's Maker than you—for
> I have sought His laws in *my* way, all my life, while you, I understand,
> have been collecting butterflies. (*The Island of Doctor Moreau*, p. 55)

It's taken directly from Bernard, who decried natural history as "passive science," distinctly inferior to experimental work that directly contributed to medical practice.

The outlook is by no means rare today and, if delivered to a colleague in such tone and terms, might prompt a return punch in the nose—largely because the modern naturalist would be most stung by the fact that outside listeners would frequently agree. Despite its extreme relevance to human life and society, ecological work struggles for a fraction of comparative recognition. This is a long-standing subcultural rift and intellectually just as unfair back then—for example, to Wallace's groundbreaking biogeographical work—but Moreau wields his upper hand in it effortlessly.

The Opponent's Key Weakness

Moreau includes evolution in his scientific view, but again characteristically for his subculture, he cherry-picks the concept of humans as animals for purposes of comparisons and proxy research while ignoring its other implications about humans.[6] He also cites its text most specifically to Prendick in the sense that "even your evolution" agrees with him—and in doing so, he exploits one of the weakest elements of the theory as it stood: its progressive interpretation as a refiner of rough systems based on utility.

> I never yet heard of a useless thing that was not ground out of existence
> by evolution sooner or later. (*The Island of Doctor Moreau*, p. 55)

The role of use or disuse in evolution came from Lamarck's work, and it became an ongoing motif as the ideas developed further, even when it fit poorly. Whenever Darwin writes of utility, or appearance or disappearance of a feature based on use or disuse, his text struggles terribly, and the theory during the following century didn't do much better.

Evolution, or specifically natural selection, does not occur "for the benefit of the species," in any way, and reduced or lost organs aren't a matter of "need" but are instead a case-by-case phenomenon depending on local variables. Consider as many species with a radical ecological shift in their history as you like, with some reduction in form and function in some body part. You'll find that plenty of these organs remain very similar to their prior form with minimal use in the new lifestyle, like sea otters' hind legs; others are present but with much of their capacity gone, like blind eyes in some cave fish; and many are gone or very nearly gone, like the leg limbs of whales.

The reason for this diversity of outcomes is that each case presents its own interplay among (i) the energy it costs to make that organ, (ii) the degree to which its reduction is even genetically available, and (iii) whether that cost interferes with reproductive effort in this new environment. If the developmental cost (i) is high relative to the energy budget, if the genetic variation (ii) is present, and if the reproductive cost is high, then selection kicks in, reducing that organ's presence in the body. But that applies only insofar as the math holds among the three things. If the selection isn't strong, then the organ remains despite its potential to be reduced; or if the selection is as strong as you like, but the genetic variation isn't there, then the organ remains; or if the selection occurs as I described but one or another cost is alleviated before the organ is gone; and so on—any situation that doesn't include all three components results in the organ remaining or partially remaining. Theoretically, the entire range from presence to absence should be observed when you look across species, which is confirmed in reality. There is no role for getting rid of an unneeded thing in the engineering sense, or finding some ideal body type "for" that particular new environment, or any goal-directed concept at all.[7]

Prendick can hardly be expected to reply in this way, though, as evolutionary theory didn't begin to address this problem until the late 1970s. Moreau is strategically using this point as it stood then, such that challenging it as a Darwinian would be undermining one's own position, and doing so quickly enough to keep

Prendick from retorting that Moreau should choose between invoking Darwinian thought and dismissing it as so much butterfly watching. I rather admire Moreau's expertise at delivering this precise flaw in current evolutionary theory as a rhetorical tactic.

The Shared Foundation of Exceptionalism

By now you might be getting the idea that I think Moreau is right. I don't, and the reason is easy. To refute Moreau, I don't need either unshakable faith in my own morality or the benefit of a century's science. All I need is not to hold an exceptional view of humans, and to call him out on that very point, when he says:

> Each time I dip a living creature into the bath of burning pain, I say: this time I will burn out all the animal, this time I will make a rational creature of my own. (*The Island of Doctor Moreau*, p. 59)

Wait, what happened to our abstract interest in the plasticity of shape? What's this about getting rid of animality to result in a new creature that we know is in fact an animal? "Come off it, Moreau! These are creatures like yourself and this is torture, as they feel it and as you'd feel it too. I don't see you avoiding pain-receptors with your scalpel—are you claiming that experiencing the pain is itself accomplishing an experimental goal?" And even more so, "Come off it, Moreau! You think you're making something special, but humans are nothing special—we're already a species of animal, which is why your re-shaping works. This has nothing to do with 'burning out' anything. Your whole goal is a mystical phantasm, and you call yourself a scientist." But to make these arguments means tossing one's own exalted status as a human out the window, for good.

The trouble with a material worldview, specifically a non-exceptionalist one, is that you have to go all the way—you can't claim cosmic and natural justification for your own position either. Supporting your position becomes much harder than merely grabbing for the brass ring of such justification harder than your opponent. In this case, Prendick can't refute Moreau about pain unless he equates their pain to humans', not merely as like it but actually the very same thing, not intrinsically worth less consideration than human pain. He can't refute Moreau about science's worth, and what kind of science, unless he puts aside the entitled assumption that science is supposed to benefit humanity uniquely. Therefore, jarringly, and only because Prendick shares the assumed special status of humanity, he gives in, more in despair than in defeat. He doesn't quite know why he's lost.

Exceptionalism lurks throughout late nineteenth-century evolutionary theory, again and again touching upon the flat observation that everything about humans was and would be a subset of animal variations, and just as often shying away because the cultural narrative demands that humans must transcend animals in some key way. Moreau utilizes it for armor in the debate. He strategically does not mention Darwin's research on animal emotions and expressions, or thoughts on human origins, which are profoundly non-exceptionalist. It leaves him open to counterargument on his vulnerable point, his complete lack of science on the single crucial issue: What is, in material observation, a human person?

As an intellectual, Prendick could combat superstition or institutional privilege, but unless he flatly refutes the presumption of God's favor toward humans as evident in His works, he cannot challenge Moreau's position.[8] It's an exquisite tension into which Moreau has maneuvered him. If he'd stuck with identifying humans as animals, period, de-nobilizing human aspirations, he'd have won the debate. But as a reformer-idealist of his time, he cannot. Therefore absent anti-intellectual, anti-abomination wrath, all Prendick has going for him is a softer heart.

Prendick intellectually and morally fails to refute Moreau in words, so from this point forward in the story, refuting Moreau must be left to action. But it's not *Prendick's* action, as I'll discuss in Chapter 6.

Pain Is Real

The novel's depiction of the historical debate was borne out for at least half a century, and its legacy is still with us. It'd be nice to say that researchers immediately switched to pain alleviation as soon as the technology was available, but they didn't. My candidate for the most egregious work would be the interventions upon and general treatment of uncountable rhesus macaque monkeys until the late 1970s, horrors that Bernard himself might have fled from.

One Life with Labs

One can read all the journal articles and books there are, but if my students are any indication, the popular discourse has been trapped in amber, with the older dichotomy still available on tap to recapitulate the Brown Dog controversy in painful detail at any time. The two windows are the default throughout the larger culture, and all the intellectual musing in the world doesn't change a cultural code this strong. In this construction, the scientific use of animals is especially painful and especially wrong. It doesn't budge in its view that both scientists' psychology and their experimental techniques are by definition *uniquely and distinctly suited* to the mistreatment of animals. "So what if you have all these rules, you're torturing them anyway because *that's what you do*." Some sectors of animal care activism have retained the word "vivisectionist" to express this point; nor is this outlook restricted to them.

The Moreau films consistently present the image—coded as normative—that no scientist is even barely trustworthy regarding the treatment of animals. *Island of Lost Souls* is the primary example, but there's no matching the scene in *The Island of Doctor Moreau* (1977) in which Moreau, presented with a subject whose behavior fails to match his performance standard, whips him mercilessly for no imaginable reason.

Movies in general blatantly depict scientists casually and habitually torturing animals, as in *The Secret of NIMH, 12 Monkeys*, and *28 Days Later*, and the list goes on. Existing legislation and standards are either ignored or dismissed. Films are particularly misleading about repeated use, depicting animals as living miserably for years in laboratories, subjected to experiment after experiment—a phenomenon I have never seen, nor would it be tolerated by any scientist I've ever met.

For context, the amended Cruelty to Animals Act of 1876 in the United Kingdom was emulated by a number of European nations within a few years, and was revised and given stronger content after a Second Royal Commission's verdict in 1912. In the United States, the Animal Welfare Institute was founded in 1951 to find common ground between researchers and activists, and in 1959 William Russell and Rex Burch published *The Principles of Humane Experimental Technique,* fairly regarded as the long-overdue rebuttal to Bernard's infamous position statement. Not only were the ethics brought forward as a valid component of science—as a human activity—but also the crucial point that pain compromises the scientific results much more than anesthesia does. These points informed the first US Animal Care Panel in 1960, providing the first *Guide for the Care and Use of Animals* in 1963, and the founding of the independent Association for Assessment and Accreditation for Laboratory Animal Care International (AAALAC) in 1965.

Federal legislation began with events similar to the fictional trigger that resulted in Moreau's defrocking, including video footage of wretched and emaciated dogs in facilities that sold them to research laboratories. Articles in *Sports Illustrated* (1965) and *Life* (1966) sparked an unprecedented, tremendous lobbying effort, which resulted in the first version of the Animal Welfare Act being passed in the same year, soon followed by the Horse Protection Act (1970) and the Marine Mammal Protection Act (1972). For the kind of research most relevant here, typically funded by the National Institutes of Health (NIH) and the Centers for Disease Control (CDC), the governing body established by the Act is the Office of Laboratory Animal Welfare (OLAW), which oversees inspections. Little or none of this legislation criminalizes the scientific mistreatment of animals; instead, their care lives or dies in the operations of each Institutional Animal Care and Use Committee (IACUC), an in-house committee that reviews all the projects using animals in a given institution (like an academic department) and is empowered to close the research, temporarily or permanently, based on their findings. In addition to the in-house members, it has to include at least one veterinarian and one member from outside the institution.

During the early period of legislation, it's impossible to generalize about the actual change in scientific practices—here and there, yes, in different ways, sometimes not at all, as with the rhesus monkeys in the Department of Defense. The big changes may have been generational, as younger scientists phased into leadership positions.

Both the Act and the Guide have been revised several times, including the publication of *US Government Principles for the Care and Utilization of Vertebrate Animals Used in Testing, Research, and Training* in 1985, providing standards for the mandated IACUCs, historically in parallel with the revisions to the Cruelty to Animals Act in the United Kingdom in the Animals (Scientific Procedures) Act of 1986.[9]

My research experiences track the reform transition to the letter. At that moment in the mid-1980s, as the national research culture shook into new form with the presence of IACUC, I became an apprentice to the world of *Rattus norvegicus*. I witnessed the local decision to apply the Act's standards to rodents despite their not being officially included, and was relieved to find that all the lab animals used in an experiment were killed quickly, that the surgical interventions were anesthetized, and that the animals were monitored for normal function. However, the trip one took to get to our rodent lab space, through facilities housing animals for multiple other projects, revealed that the macaques still lived in bare metal cages—a practice soon to be discontinued and regarded by my generation of scientists with loathing.

My research experience also includes museum collections, which are immense deposits of body parts, most of the animals having been killed toward that end, or, in that subdiscipline's own euphemism, "collected." This work is arcane even to other biologists, but it's crucial, literally grounding our knowledge of life's diversity in geography, ecology, and verifiable specimens. Without it, all other biological investigation loses its relevance—even our basic knowledge of how many species exist of any given type of living creature. The ethics issues in this work are also profound, especially since they often must be applied in a complicated web of different nations' laws, and require a lot of decisions about death and pain on the fly.

Within science, a wide range of independent dialogues had begun, indirectly informed by the activist world, but focused on science specifically—not, for instance, the meat industry, product testing, or performing animals. The discussions multiplied and started jumping disciplinary boundaries to produce to new symposia and debates. One early paper (1989) by Strachan Donnelley, "Speculative Philosophy, the Troubled Middle, and the Ethics of Animal Experimentation," described scientists as a "troubled middle between human welfarists and animal rightists," which I believe was the first time the two-windows problem (not by that name) was identified. Some scientists knew they were not brutal torturers of animals in the name of human benefit, or if they were, they wanted to stop and do it a different way—but they also knew that simply switching windows was no solution at all. Donnelley suggested that such scientists address the problem themselves without necessarily buying into what he described as the "ethical three-ring circus" of animal activism ideologies.

A few years later, the journal in which he published, the *Hastings Center Report*, would publish the proceedings of the Sundowner Conference, held in 1996, which yielded a set of principles similar to the Belmont Report concerning research on humans in 1978–1979. These and other results from this period have produced a significant library with a voice of its own, appearing in book form in the early 2000s.

This drive in internal reform was paralleled by external activism, and by the late 1980s, the leading activist writers, topics, and orientations had fallen into identifiable lines of thought, as follows:

- Utilitarian: most identified with Peter Singer (*Animal Liberation*), most applied toward with animal welfare, originally introduced with the view that the current level of inflicted pain is not worth the benefit, associated with improving treatment while reducing, not necessarily eliminating, use.
- Rights: most identified with Tom Regan (*The Case for Animal Rights*), mostly applied toward reducing use as such.
- Combination: most identified with Richard Ryder (*Victims of Science*), including the widely adopted term "speciesism"; here the most relevant concept is his term "painism," combining the utilitarian view that pain must be reduced with the rights view that animal use is questionable—or outright wrong—even in the absence of pain.

This construction doesn't mesh neatly with the internal scientific reform that was under way. Conservation work, for instance, had established that sentimentalism about single individuals in captivity or about single species didn't result in effective policy, so scientists killed and collected animals in the field toward the goal of understanding and preserving the entire habitat—often successfully. Physiological and behavioral work in laboratories underwent many different changes based on institutions and on the specific animal, and my own research experiences fell right into the transition between imposed standards and internalized reform.

I discovered the limitations in ethical research and teaching the hard way, losing half of the mice I'd field collected to mishandling at the post office, objecting to the treatment of turtles in the physiology class, and losing one animal in my vole colony, after over a year of successful care, to an aggressive cage-mate.

I canceled that project in full, for which a certain sentence in the novel seems resonant:

> Then we went into the laboratory and put an end to all we found living there. (*The Island of Doctor Moreau*, p. 82)

My eventual work made use of existing museum collections rather than live-animal work. I'd found a pretty solid spot for my own ethics. Bluntly, I was an untroubled killer, either directly or indirectly, but not a willing agent of pain, insofar as research standards were able to ascertain. It also seemed right to be honest about it, and I resisted using the standard lab euphemism of "sacrifice," which I still think is a weird thing to say. My line appeared at collateral damage. Deaths that I planned and conditions I monitored were one thing, but conditions that yielded even a little accidental death and agony along the way were another. To do otherwise would be "to shoot and cry about it," a position I refused to take.

All scientists I knew who worked with live animals did the same thing: they self-regulated, finding a creature, a topic, and techniques inside their personal limits, and those were typically already inside the larger institutional limits. The problem with this, however, is that it also caused them to *self-censor*, resulting in no ongoing dialogue and no sense of community identity regarding animal care. Anything beyond one's personal processing was left to comparatively sterile concerns like "enough to keep my job," and that context quickly develops cracks. That why the biggest problems weren't in the research intervention but in the more general, day-to-day husbandry. The period for my own research with live animals corresponded with this escalation in dialogue, although the institutional improvements weren't yet in place. Although we graduate students were generally animal lovers already, trained by animal care staff, and technically overseen by advisors, in the moment we were often on our own in terms of engagement with the regulations or ethics—especially those like me who did original work rather than filling an existing slot in an advisor's big research project. We didn't need ethical consciousness-raising, we needed robust policy.

In 2001, the IACUC at the Children's Memorial Research Center (CMRC) invited me to be their outside member, bringing me back in contact with live animal work after about ten years' absence. The research topics at the center were right at the cutting edge for some of the most challenging topics in applied biology: stem cells, cancer, neural development, regeneration, and gene regulation, mostly associated with natal problems or inherited conditions. If you wanted to know what biotechnology could do, or might do, or was under argument for what it was doing, then this was the place.[10]

Nothing I'd seen compared to an animal-use endeavor like this one. The animals there were almost all rats and mice. The total rodent housing capacity was about 20,000, if they were mice (figure half that for rats), but at no time was this capacity approached.

For a while, rabbits were used as training surgery animals for classes on infant surgery, and some of the facilities were prepared for larger animals like pigs, but no larger animals were used during that time. The rats and mice were mainly housed in a centralized colony, meaning banks of plexiglass cages, which supplied the animals to individual researchers and whose staff monitored them, whether in each lab or in the main colony, and whose supervisor was also a veterinarian and a committee member. This is a bit different from my experience at universities, in which a given lab conducts its own animal husbandry, and it made the committee's job a lot easier.

I hadn't seen live rodent work for almost ten years, and the changes in practice were stunning. The biggest funder of biological research in the United States, the National Institutes of Health, was now accredited by AAALAC, which means its funding has to follow much more stringent mandates than those of the Animal Welfare Act. Although a motion to include lab-bred rodents' and birds' coverage in the Act failed in 2002, treating these animals in accord with the "animal" definition in the Act was already the standard practice anyway (not that it would have hurt to see it made official in the Act).

In *Responsible Conduct of Research* (2003), biochemist Adil E. Shamoo and philosopher David B. Resnik provide the detailed history, and it was a long road, longer than it should have been in my view, with some bad moves along the way. But the tipping point seems to have arrived. The numbers of animals used in research was reduced by a full 50% from the 1970s to the 2000s, according to both the US Department of Agriculture and the Humane Society. The Great Ape Project reported in 2005 that 3100 nonhuman primates were kept in research facilities in the United States, with 2180 used in research, which is infinitesimal compared to even a decade before. Education in animal care history and ethics is now mandated content for graduate studies in most research programs.

The required in-house committee overseeing animal research, IACUC, is now expected to enforce three mandates:

- Animal care applies not only to animals' direct use in experiments but also to their husbandry, or living conditions, which means that it applies at all times to any and every animal in the scientist's or institution's control.
- All animals used in any research project must be killed at the project's end. Not only does this account for possible trauma or effects that being in the study might induce, it ensures that no animal is used in more than one study.
- Stress and misery are considered equivalent to outright pain, and are subject to every rule concerning pain.

Per project, the rubric is explicit: a given study must minimize the number of animals used, prevent or minimize pain and stress, and address an understandable, relevant question. The five categories of pain, A through E, are worth a good look because A, the "least," is never assigned—by definition, being subject to human control already puts the animal into B.

No project could be done at all pending committee review. We reviewed each one at its initiation and on a following schedule, usually going over five or six projects per monthly meeting. The committee also oversaw the animal housing for the general colony and in each of the labs, when the latter had any, including the numbers present and how many were being used, and including a walk-through of the whole facility every six months. Not to put a fine point on it, if the work—funded as it might be—didn't pass the committee, it would be suspended or shut down.

If you think that the committee's composition of mostly researchers from the facility meant taking it easy on one another's colleagues, think again. We were strongly influenced by Shamoo and Resnick's book and did not consider the work to be mere compliance. The standards were set as follows:

- Most broadly by the Animal Welfare Act
  - Sharply limited by the standards of a third-party agency in the later years
  - When applicable, limited by the standards of a granting agency
    - Specified by our own research of the techniques, especially monitoring the latest knowledge by the veterinarians
    - Specified to the specific techniques and practices of a given researcher as known by us directly.

The CMRC consistently passed external inspections, and while I was there the institution applied for accreditation with AAALAC, receiving it in 2008. We paid special attention to pain categories; to endpoints, which is when and how subjects are scheduled to die; to standards for monitoring animals for evidence of misery or stress; to the research literature to assess relevance and redundancy; and to the latest information on pain-alleviating substances.

My part often involved critiquing a project's experimental design to see if it really addressed the stated question, so that more animals would not be suddenly requested halfway through the project, or that some category provided with animals wouldn't turn out to be unnecessary. A given researcher and I were often able to design away from multiple interventions per animal, which as it happened was the substantive point regarding Bayliss's demonstration of the famous Brown Dog incident and the subsequent trial in 1903. Sharpening up the criteria for experimental design meant arriving at the most honest animal numbers possible.

I stress this because utility is the weakest requirement in the IACUC mandate and in the entire discussion of research animal welfare. I don't care how glowing one's proposal for funding is; no one ever knows whether a whole sector of research will bring "utility" or "benefit," or when, or even what that is, let alone for a single project. Demanding an immediate benefit militates against basic research, the wellspring of ideas in science, and therefore reduces potential research benefits considerably. Instead of effectively forcing people to lie about how their work would usher in undreamed-of benefits, we required that the experimental designs make maximum sense, that a study had an embedded "yes" or "no," or genuinely forged into the unknown.

The committee did sometimes take punitive action. If animals in a researcher's care weren't following the standards to the letter, for instance if the animals' cages weren't being labeled completely, we'd cite them with the explicit threat that they'd lose their access to animals unless it was corrected. Most of these instances arose from a technician's momentary carelessness, and the researcher fixed it immediately. Some projects were simply denied, usually from external researchers or add-ins that hadn't originated at the CMRC, for example a proposal to work with chimpanzees in one case, which simply had no reason for such a subject that we could see.

In three cases that I can remember, we halted a current project and closed access to the animals; the most serious infraction involved a cage of mice left on a car seat for a while, and although they came to no harm, it was considered a startling lapse. Not once did I observe a case comparable to the treatment of lab animals prior to the Animal Welfare Act, nor did any researcher ever dismiss animal care as a viable issue for his or her work.

I can't speak to the larger topic of research as a whole, because there isn't any such whole. Mistreatment of animals was not a myth in the 1980s, as exposed by activist films such as *Unnecessary Fuss*, and I can't claim that my little life history in science represents the most powerful social force in the gains made. I do know that I'd aged into a time and position in which we scientists collectively meant it, and we considered care to be a work in progress—which needed progress—rather than abiding by perceived industry standards. These values didn't arise under direct pressure from activism, but from our collective experiences. I think that the activism had its effect more subtly than its major advocates perhaps wanted, meaning that we'd all grown up with it and thought about it for ourselves. Several of the researchers and administrators had been active in animal welfare and stayed in research—in this case, very ambitious research—rather than condemning it. We regarded pain, stress, and misery as a constant danger because this is animal use, period, and considered ourselves not just as a committee, but as part of the community, to be—or rather, do our best as—moral agents. In that, Strachan Donnelly had been right on the mark.

Many activists for the reform of animal care are to be credited for their understanding of this history, especially Peter Singer, and I also cite Rod Preece, who, although hardly supportive of animal research, does not demonize it and clarifies in his *Animal Ethics* that no human culture is especially nice to animals despite a tendency to make such myths about cultures outside one's own. Since the 1970s, some branches of animal welfare activism have found a positive interlocking path with the generational changes in scientific culture.

I want to focus on the two-windows problem, though—not only is it still present in branches of animal activism, it's also—as I said earlier—powerfully entrenched in the larger culture. *The Island of Doctor Moreau* shows the way, right in the breaking point between Moreau and Prendick: human exceptionalism. Much activist discourse is still stuck in it: the idea is that some creatures are afforded privilege against suffering—with ourselves as the gold standard—and the question is which others are permitted this privilege—as judged and implemented by ourselves. The relevant texts are startlingly frank about "lower" animals and about eligibility based on their similarity to us, or how much they can suffer when compared to our assumed maximum capacity.

Other branches of activism seem disconnected from biological information and positions about nonhuman behavior. After over 300 pages of multiple authors' point-counterpoint debate in *Animal Rights,* including all manner of abstractions, Martha Nussbaum's conclusions concede that some research is allowable, but that animal rights demand that ecological habitats should be preserved and that animals utilized in experiments should receive dignified deaths. I'm not certain what scientists she has consulted with, but possibly not very many. Habitat preservation and painless euthanasia already receive unanimous scientific support, and have for decades.

Look at Moreau, because in this, he's exactly right. He's a not a villain in a "don't meddle" story, as he's not trying to emulate or outdo evolution, not trying to be anyone (or Anyone) he's not, not seeking fame, and not whining about being laughed at "back in the academy." But more important, he's not the vivisectionist imagined in the utility/pain dichotomy either, because he's not bragging about the inconceivable benefits his work is about to bring. He simply wants to know; the problem is not a utilitarian one, but a matter of how much pain he's willing to inflict. Moreau is completely aware that the subjects suffer; telling him so in hopes of him suddenly recanting is futile. What's painism when you're willing to do it? What are designated rights to the powerless, or to some subsector of the designated rights-bearer from whom they're withheld?

We're supposed to be exalted Man and we can't arrive at an ethic for inflicting and managing suffering? The ethics discourse keeps running up and down the arrows between the levels of the Man/Beast dichotomy. Moreau's and Prendick's debate is right on target, in exposing human exceptionalism as the culprit, the real intellectual agent behind the two windows. Once humans are neither masters of the lowly beasts nor magnanimous uplift-agents for (some) beasts to be included, the three-ring circus merely disappears.

I was technically part of Donnelley's "troubled middle," with the minor correction that I—as with many of my age group—were not precisely troubled, which would imply that we were confused. We knew exactly where our limits were; we'd found them ourselves at various crises within our work and even at costs to our careers. When we could, we took the helm. It was a matter of pure practice, ahead of formal policy rather than trying to legislate it and then force compliance. It also bypassed the debate about some incontrovertible "ethical good" in isolation. We knew the issue is not pain traded off with utility. It is instead the interplay of knowledge, care (or mercy, as Christopher Scully calls it in *Dominion*), and death, or more accurately, killing. The nearest to a formal voice is found in Nikola Biller-Andorno concept that animals' capacity to be harmed or, more generally, misused is a matter of no controversy among scientists. Our question instead is whether and when we can harm them, and framed that way—and acknowledging the role of activism as pressure—I think it resulted in more reform than in any other period in scientific history.

Readings

Readable summaries of medical developments during the nineteenth century include Michael T. Kennedy, *A Brief History of Disease, Science and Medicine* (2009); Roy Porter, *Blood and Guts: A Short History of Medicine* (2004), and focusing on animal subjects, Hilda Kean, *Animal Rights: Political and Social Change in Britain since 1800* (1998). Extensive biographical information about Claude Bernard and many of his original papers are available at the Bernard archives (claude-bernard.co.uk); and his *An Introduction to the Study of Experimental Medicine* (1865) is an education in itself.

The difficult and politically charged concepts of utility, progress, and biological processes during the nineteenth century are explained in Robert J. Richards, *The Meaning of Evolution* (1993).

Current and past standards for animal care in research in the United States are all available, through the Public Health Services, including the *Guide for the Care and Use of Animals* (8th edition, 2010) and its associated standards for IACUC work. All these should be read in the context of William Russell and Rex Burch, *The Principles of Humane Experimental Technique*, from 1959, and Strachan Donnelly's "Speculative Philosophy, the Troubled Middle, and the Ethics of Animal Experimentation," *Hastings Center Report* 19(2), 1989.

Books like *The Monkey Wars* (1995) by Deborah Blum; Bryan Norton, Michael Hutchins, Elizabeth Stevens, Terry Maple, *Ethics on the Ark* (1996); F. Barbara Orlans, *In the Name of Science: Issues in Responsible Animal Experimentation* (1996); and Phillip Iannaccone and D. G. Scarpelli (editors), *Biological Aspects of Disease: Contributions from Animal Models* (1997) capture the diversity of views held by the incoming generations in science in the 1990s.

The many works over many years by Peter Singer and Rod Preece are a library of their own, all well known and available for investigation, and more overlap may be found with the internal ethics of modern science than one expects. My topic does not include direct-action groups like the Animal Liberation Front or the Animal Rights Militia, or specific social organizing like People for the Ethical Treatment of Animals, but certainly a broader discussion of policy would.

It's useful to compare Lawson Tait's departure from research, and his *Last Speech on Vivisection*, delivered to the London Anti-Vivisection Society in 1899, with the similar experience of Richard Ryder, whose works include *Speciesism* (1974), *Victims of Science* (1975), *Animal Revolution* (1989), and *Painism* (2001).

Cass R. Sunstein and Martha C. Nussbaum (editors), *Animal Rights: Current Debates and New Directions* (2005), features point-counterpoint essays by many prominent writers in animal welfare and animal rights activism.

The newer library of animal care is large, but good starting points include John P. Gluck, Tony DiPasquale, and F. Barbara Orleans (editors), *Applied Ethics in Animal Research* (2002); Adil E. Shamoo and David B. Resnik, *Responsible Conduct of Research* (2003); and Nell Kriesberg's *Research Ethics Modules* (http://www.ncsu.edu/grad/preparing-future-leaders/rcr/modules/index.php)for North Carolina State University, especially "Animals, Science, and Society." For an understandably less conciliatory view, see Dario Ringach, "The Use of Nonhuman Animals in Biomedical Research," *The American Journal of Medical Research* 342(4), 2011.

Notes

1. The women Sims operated on were slaves, which blurs the distinctions among human patients, animal patients, captive animals, human prisoners, and experimental subjects.
2. Claude Bernard, *An Introduction to the Study of Experimental Medicine*, 1865, translated by Henry Copley Greene (1927), published by Dover Books Inc., 1957.
3. Bernard argued against the use of summary measurements, which seems strange today, but his era of research preceded methods to assess variation. Lacking the distinction between standard deviation and standard error, his criticism was valid. Modern inferential statistics combine such methods with the rest of Bernard's points.
4. Alexander Comte is a key figure in the development of today's political and ethical vocabulary, not least in adopting evolutionary terms into progressivism, such that today policies and persons are still spoken of as "evolved." He also coined the terms "secular humanism" and "altruism," both of which carry implications of really human, not animal, as virtuous behavior. As I understand these matters, David Hume and Arthur Schopenhauer had demolished this position before Comte popularized it.
5. In "Claude Bernard and An Introduction to the Study of Experimental Medicine: 'Physical Vitalism' Dialectic, and Epistemology," *Journal of the History of Medicine and Allied Sciences* 62(4): 495–528, 2007. Sebastian Normandin suggests that Bernard's research was vitalistic. I am not so sure. The initial chapters in his Introduction use "vital" as a synonym for "alive," and in citing its causes as strictly physical-chemical, seems to be in the classic Lawrence mode. But in Part II, he claims that "the vital idea" directs embryonic development, while preserving (material) determinism for the means. Further passages slip back and forth between the two meanings. I bring this up because the question of Moreau's materialism or vitalism is subtle. His one fixed idea about the rational man born from a bath of pain is indeed more like a medieval alchemist than a modern scientist, practicing

a transmutation of the flesh to accomplish a spiritual goal, but that idea stands in sharp contrast to all his other views. If he's a vitalist, then he's a later sort, not so much about magic forces but using rationality for his distinguisher, a presumed freedom from urges and especially from fear.

6. Bernard does not mention natural selection or Darwin, but he does acknowledge evolution, like all nineteenth-century scientists and scholars of whom I'm aware. Hugh LaFollette and Niall Shanks are mistaken in stating that he did not ("Animal Experimentation: The Legacy of Claude Bernard," *International Studies in the Philosophy of Science* 8(3): 195–210, 1994), as well as that he misreads comparative studies; his logic is perfectly sound.

7. Vestigial organs also cause difficulty in the modern discussion of evolution, because one might ask, if evolution discards unneeded things, why not discard them entirely? It also raises impossible questions about whether an existing organ can be called useless insofar as it's part of a functioning body. The better argument in favor of evolution is to consider all organs in terms of homology, regardless of their current operations and activities, and therefore less-active organs become useful road maps of historical change, with no reference to how "useful" they are.

8. The dividing line between sentimental anthropomorphism and a recognition of similar experiences and suffering is difficult given our current vocabulary. Sherryl Vint's *Animal Alterity* (2013) and Nicole Anderson's many journal articles provide some new terms that can help, especially when discussing pain and quality-of-life standards in research based on the animals' own experience.

9. Animal welfare and rights activism is often tied to other current issues, such as environmentalism and opposition to the Vietnam War in the 1970s. Members of ALF and ELF today claim historical solidarity with activist groups like Weather and Earth First!, with a characteristically difficult overlap with science-based activism like global warming and stem cell research.

10. When I joined the committee, the institution was named the Children's Memorial Research Center, then the Ann and Robert H. Lurie Children's Hospital of Chicago, then the Children's Memorial Institute of Research and Education, and most recently the Stanley Monroe Children's Research Institute. It is affiliated with Northwestern University but is technically freestanding based on funding from its governing agency, the Lurie Foundation. All research positions there are also faculty at Northwestern, and it follows university rather than private standards for training.

5

Into the Lab and onto the Slab

Remember what Moreau says, as quoted in the previous chapter:

> I wanted—it was the only thing I wanted—to find the extreme limits of plasticity in a living shape. (*The Island of Doctor Moreau*, p. 56)

All of biology resides in this sentence, because the underlying chemistry and structure of life is all the same. What remains to be understood, as the science of the nineteenth century discovered, is its remarkable inherent dynamism. Living things not only change individually, they change as groups, and at a far larger scale, the composition of life's diversity across groups has changed. Nor is this a long-ago story; all of its dynamic processes are still in operation right now.

The hard part is that living processes are layered and nested, with different changes and effects at different levels. We have diversity and changes of chemical processes for both inheritance and physiology, then diversity and changes of body shape and functions for a given individual's life, then the diversity and changes of body shape and functions across different creatures, and finally the appearance and disappearance of types of creatures, all within the staggering contextual question of how the larger, interacting environmental cauldron of chemistry and physics imposed itself on this history.

Saying "natural selection!" isn't enough. The childish version of that idea says, "mutations are random and environments shape the outcomes," giving the false impression that *anything* can or could evolve because it might "mutate in" from the Nth dimension, and the guiding hand of the environment says yea or nay. The thinking version asks, instead, in the real world, what were the historical limits, and what are the hard limits beyond those? Are there different categories for different levels, like genes and chemicals, versus those for potential body shape and functions, versus those for environments and selective pressures? What shapes or functions cannot evolve among living things as we know them? And why not: contingent historical events, or hard-line chemical and biomechanical constraints?

Moreau's project is better than it looks—it directly examines the limits of individual shape and function across the boundaries between species. Darwin suggested that these boundaries were strictly contingent details of different selective histories, so that there really isn't anything such as a species boundary as far as causes are concerned, only effects. Was he right? He also suggested that we humans are one species among many, without special properties or historical processes. Would inquiry into species boundaries reveal the same results between us and others, as among them?

What if it were done without pain, without disregarding the experience of the subjects? *Can* that be done? The answer is—almost: yes and no, with the "no" being a matter of questions rather than brick walls. How abominable would that be?

How

What Moreau does to his subjects, effectively slicing and dicing them into new anatomical shapes, wouldn't work. The pain involved would probably kill his subjects through shock, for one thing. The principles of its fictional effectiveness, though, make more sense than one might think. The terms chosen for his explanations are quite specific and are traceable to presentations and experiments made in physiological research during the mid-nineteenth century. For example, his phrase "the physiology, the chemical rhythm," is a direct and accurate reference to Bernard's *milieu intérieur*, prefiguring later discovered details such as negative feedback regulation and set points. His discussion of emotional functions being "seated" in the brain are nearly equally directly from the work of the historical Moreau, although it oddly does not include using psychoactive substances, as those would seem to fit right into our Moreau's techniques.

The one thing that doesn't fit at all is his concept of pain as transformation, because it's bonkers, as I discussed in the previous chapter. That's a metaphysical issue for him, perhaps even unconscious, as he only mentions it when excited, and it contradicts his earlier cool dismissal of pain as irrelevant to his goals. Fortunately that whole thing can be set aside.

A lot of Moreau's techniques rely on damaging tissue and controlling the gross physical circumstances of its healing. Many elective surgeries, like nose jobs, involve revising a given feature's appearance through such methods. The urethro-vesicular correction I mentioned in Chapter 4 is a therapeutic example. An extreme version is to transplant tissue from one part of the body to another, as with skin grafts, and in many cases, this works well. Sometimes, the regrowing organs can be astonishingly flexible—for instance, if you amputate a limb, say a leg at the knee, you don't do anything to the internal anatomy but merely fold over the skin to make a stump. But inside, the capillaries at that point will regrow to connect the arteries leading to that location with the veins leading back from it, even though before the amputation, they didn't meet there at all.

However, not all tissue is as flexible, based on what type it is and what's being done to it. Sometimes it can't repair itself well, or at all, as with most of our nervous system, and in other cases, what you've damaged may well repair as itself, not as what you tried to change it into.

Moreau's techniques also include blood transfusions and, extensively, tissue transplants across species, which runs right into trouble. Both sets of techniques are facing the considerable array of the recipient's immune system, which has no idea that this incredible invasion of alien cells is supposed to be therapeutic, and responds so aggressively that all sorts of negative effects occur. At the time of writing, the proteins that define human blood type were not quite yet understood. Until 1901, no one knew why providing a person with someone else's blood sometimes worked perfectly but sometimes lethally did not. Once the proteins that prompt the immune response were understood better (and fortunately the blood types are so biologically simple and individually distinctive), the technology became reliable over the next two decades. Transplant rejection was even less understood, to the extent that people may have thought its failure was a matter of insufficiently refined surgery, rather than a tissue response. I shudder to consider that even through the 1920s and 1930s, people like John Brinkley and Sergei Abrahamovitch Voronoff were still hawking transplants of, respectively, goat gonads and primate gonad/thyroid tissues into humans. The first successful human organ transplant from a living donor was in 1954, between identical twins, soon followed by immunosuppressant techniques that enabled a person to receive organs from a non-identical donor.

Certainly all the cross-species transplants, which play a big role in characterizing the various Beast Folk, were impossible. However, none of them was intrinsically fantastic or silly in terms of one-step-into-the-future science fiction.

What Are We Asking?

Let's look at the explicit question, whether humanity can be coaxed to arise from nonhuman animals, and the first step is to acknowledge is that this is a biologically incoherent statement, and to rephrase it sensibly: whether a functioning member of *Homo sapiens* can be developed from the zygote of another species.

The discussion of assessing our measured variables all resides within a bigger discussion of experimental design: What is being compared to what, and what is being asked? Answering these questions is easy if you break the work into three distinct groups: the experimental subjects, the positive controls, and the negative controls.

Who are the experimental subjects? What species will provide these beginning embryos we are talking about? If the techniques have any *a priori* chance of working (and if they don't, then we should scrap the project until they do), then arguably the easiest would be another ape. This leads to the interesting issue of whether that would be *so* easy that it would be cheating. Is that the question, whether human posture and various other functions can be developmentally produced from those in such closely related creatures?

That better not be the question. That's a boring question. The question as I see it must be, to what extent is the incredible *specificity* of a given mammal the outcome of developmental rules that *all mammals* share? Mammals are pretty incredible when it comes to tuning a limited number of genes and tissues to very different forms. Why here, see exactly what the whole Evo Devo revolution has been about:

> . . . are the morphologies corresponding to empty spaces defined by my data readily accessible developmentally?

> To address [*this family of*] questions, three separate strategies of the developmental approach to adaptation (embryology, manipulation and comparative studies) have sprung up, and making explicit the connections between them gives clear avenues for future work. Each strategy offers distinct advantages. The comparative approach offers generality, with inferences spanning many species, and is based on the products of natural, not laboratory, evolution. Studies of ontogeny of single species provide mechanistic detail, and manipulation methods can examine performance or fitness of variants directly. All are means to explore developmental potential in an adaptive context, to determine what is common, what is possible, and what is prohibited. Deliberately combining strategies would fortify inferences by drawing on the strengths of all three. (Mark E. Olson, "The Developmental Renaissance in Adaptationism," p. 285)[1]

That is Moreau speaking, or rather, it's Moreau the scientist rather than the pain-transformer and romantic admirer of the human mind. Our project, then, is to do just this: ontogeny + comparative + experimental.

For that question, the human form itself, as a species of primate, is of no special interest. The good reason to include people at all is because so much is known about our species' genetics and physiology that it jump-starts the range of necessary studies about the other. But choosing which species provides the target form must rely on more than an ineffable romance attached to our own, which is to say vanity. That was ultimately the biggest flaw in Moreau's project, as I described it in Chapter 3.[2]

The scientific question makes most sense if it studies the constraints in both directions, or in plain language, fair is fair. That means we need to see if species X can become a *Homo sapiens*, and also if *Homo sapiens* can become a species X. This design has a nice elegance, too, as the positive control for species X is the negative control for *Homo sapiens*, and vice versa.

Who is species X, then? If this were all homage to the novel, I'd include lots, first among them the black bear (*Ursus americanus*), leopard (*Panthera pardis*), puma (*Felis* or *Puma concolor*), spotted hyena (*Crocuta crocuta*), domestic pig (*Sus scrofa*), ox or rather cow (*Bos taurus*), and domestic dog (*Canis familiaris*), for reasons pertaining to the most relevant Beast People in the story. However, not only would back-and-forth crossover comparisons among multiple species be overwhelming to conduct and design, these species represent a pretty tight cluster among the diversity of mammals, all effectively the same degree of difference from primates (Figure 5.1). Much as my literary side would gravitate instantly to the puma or hyena, this has to make some kind of phylogenetic sense.

Subject to much debate, I'm sure, my call is to begin with ordinal-level comparisons within the Euarchontoglires, and among those, looking right at me, are another two species we happen to know quite a lot about, which is to say *Mus*

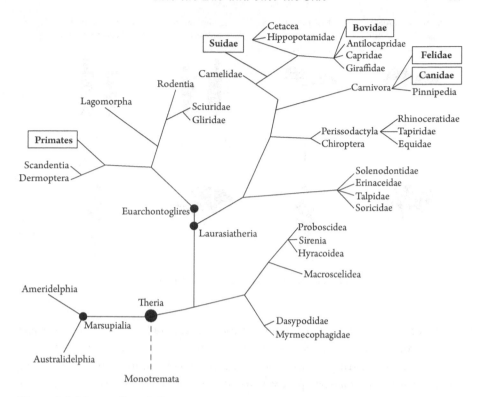

Figure 5.1 Mammalian phylogeny.

musculus, the house mouse, and my old friend *Rattus norvegicus,* the Norway rat (which is not from Norway, but Indochina; I have a personal tic about saying that). That might seem a bit of a letdown, but if you want, consider this just the beginning to a gradual stepwise comparison, branch by branch.

We have two experimental groups, one beginning with rats and ending with humans, and the other beginning with humans and ending with rats. The two control groups operate as criss-cross positive and negative controls, as in Figure 5.2:

- Control group 1 are rats that experience all the same conditions but without interventions, expected to develop into ordinary rats; they are the negative controls for the experimental humans and the positive controls for the experimental rats.
- Control group 2 are humans who experience all the same conditions but without interventions, expected to develop into ordinary humans; they are the negative controls for the experimental rats and the positive controls for the experimental humans.

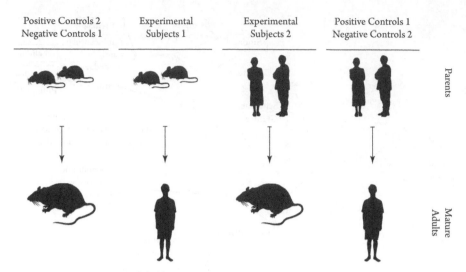

| Positive Controls 2
Negative Controls 1 | Experimental
Subjects 1 | Experimental
Subjects 2 | Positive Controls 1
Negative Controls 2 | |

Figure 5.2 Experimental design groups with people and rats.

If the control groups don't turn out as expected, that tells us something is dreadfully wrong with our entire conditional setup. If they do, then they provide a baseline for comparing and assessing our experimental individuals. A completely ideal design from some perspectives would even have the controls be clones of the respective starting species, but I think that would introduce more compromises than it would solve potential problems.

Experimental Variables

Development is the key. This will be about growing a creature, not sculpting one. However, before anyone shouts "Genetic engineering! DNA!," my first call is that we can't use human genetic material, because that dodges the issue. The textual Moreau was determined to create humans *without using human sources at all*, sticking to what the nonhuman animals provided singly or in combination. The same applies here, although I can rephrase it in more rigorous terms. Remember, this isn't an engineering experiment to get a human from a nonhuman as cheaply and easily as possible, so it's not about genes and closest-possible relatives. Instead, it's about the plasticity of development, so discovering whether genetic differences operate as a constraint is part of the point.

However, Moreau did fall right into another bad design feature: defining humanity in an ineffable, "know it when I see it" fashion, which doomed him to disappointment. I have to avoid that trap, as he did not, by setting pre-experimental design goals in strict performance terms, without recourse to abstractions like rationality, consciousness, looking into a creature's eyes to see its soul, or any other functionally vague referents based on exceptionalism. We need at least one performance variable.

Still, defining the performance goals isn't simple. There is no single feature that we can point to in isolation and say, given *this*, the creature is human regardless of anything else. When we look at someone and tag them as a fellow human, we're actually seeing quite a few things at once. To focus on the anatomical and physiological (because sadly, people also do tag others as non-human on the basis of cultural practices), these things include

- Posture: pelvic shape, thigh orientation, knee articulation, vertebrae count, lumbar curvature, spine and skull articulation, plantar foot, reduced and fused tail
- Hands: length and specific orientation of the fingers, degrees of opposition, and pressure among them
- Skull morphology
  - Expanded cranium
  - Facial and other cranial proportions, for example, ear placement
  - Reduced mandible, associated features of palate shape
- Dentition: number of each type of tooth, shape for specific teeth
- Cerebral cortex: increased volume, dense microanatomy (note: none of the rest of the brain needs to be changed, in terms of size and shape)
- Characteristic laryngeal, lingual, and other hyoid anatomy
- Eye anatomy, for example, pupil shape, iris color, retinal neurology
- Mobile facial musculature, everted lips
- Mammaries
  - Two teats positioned at the pectoral region
  - Breasts in addition to teats
- Reproductive system
  - Characteristic sizes of the penis and clitoris; absence of the penile sheath for the former; absence of baculum and baubellum
  - Simplex (single-chambered) uterus and duct anatomy
  - Location of the testes (outside the body cavity) and location of the scrotum (behind the penis)
- Specific anatomy and capabilities of the digestive tract, including a particular shape for the cecum (appendix)
- Kidney shape and specific range of water-salt regulation
- Characteristic metabolic rate (fortunately boring)
- Aseasonal reproductive capacity
- Monthly estrous cycling including true menstruation, reduced chemical signaling of fertility
- Thickened dermis, absence of underfur, specific colors, morphology, and distribution of cover fur

- Shape of the cartilage of the pinnae (outer ears) and nose
- Life history
  - Longevity
  - Growth rates and body size at all stages
  - Time to maturation.

Listing these variables raises crucial questions: to be rated as successful, how many of these features need to be made human or rat, which ones for sure, and for each one, how human does it have to be? Reproductive capacity isn't a priority, as we're discussing species differences in particular functions, not literal speciation. But aside from that, what's more important, hand anatomy and appearance, or the placement of the eyes and shape of the pupils? Would the species-specific combination of color, thickness, and length of hair be an irrelevant detail or a deal-breaker?

This is where the project needs more intellectual grounding. It is absolutely unacceptable to set the project's response variables by various recognition features hardwired into our own sensory and psychological systems. If I were reviewing a grant proposal for this project, that's the first thing I'd check out: exactly why the performance variables, and the standards they're to achieve, are deemed important.

For example, what about features of cellular and tissue function? To be successfully experimentally human, would the rat-based experimental subjects need to exhibit human-characteristic onsets, locations, and types of cancer? And of course, reversed for the human-based experimental subjects?

I also need to consider behavioral criteria, which is much trickier than it looks. "Can it talk and build a fire" is definitely not on target. It's about as rigorous as "look into its eyes to see its soul." The project must be free of some deeply rooted assumptions. For example, there is no reason to think the experimental subjects would be inherently unregulated or violent, because calling such behavior "beastly" is a value judgment, not critter-descriptive at all. Nor should they be expected to be childlike or primitive in any meaning of the word, nor should they be expected to exhibit some magical moment of "awakening" from beastlike confusion into human awareness.

During their debate in Chapter 14, Moreau and Prendick disagree about the cognitive capacities of nonhumans:

> "But," said I, "These things—these animals *talk!*"
>
> He said that was so and proceeded to point out that the possibilities of vivisection do not stop at a mere physical metamorphosis. A pig may be educated. The mental structure is even less determinate than the bodily. In our growing science of hypnotism we find the promise of a possibility of replacing old inherent instincts by new suggestions, grafting

on or replacing the inherited fixed ideas. Very much indeed of what we call moral education is such an artificial modification and perversion of instinct; pugnacity is trained into courageous self-sacrifice, and suppressed sexuality into religious emotion. And the great difference between man and monkey is in the larynx, in the incapacity to frame delicately different sound-symbols by which thought could be sustained. In this I failed to agree with him, but with a certain incivility he declined to notice my objections. He repeated that the thing was so and continued his account of his work. (*The Island of Doctor Moreau*, p. 54)

This exchange is dated and almost completely blended between obsolete and still-powerful points, and the framework of understanding is so different—cell theory was very young, development was not tied well to evolution, and no useful principles of genetics were yet discussed—that there's no point in engaging with their direct disagreement. I'll extract the valid points and frame them in modern terms.

Moreau is right about one thing. The current understanding of nonhuman behavior suggests that they—or rather, a broad and numerous plurality of separate "theys"—are already doing very many of the things we'd like to reserve for ourselves and call "human." Every animal studied for cognitive abilities can count, for instance, as long as we're talking about variables and a scale of measurement that are relevant to the creature's life-history strategy. The project doesn't need to instill so much as to permit and to redirect.

However, he's wrong about the specifics being wide open; there's no such thing as a mental blank slate. The problem lies in falsely separating mental processing into instinct (Beast) and learning (Man), which is where he and Prendick get hung up. Instead, it's better to start with the observation that developing sophisticated behavior may include learning (trial-and-error, for instance) or it may not, depending on the organism. Therefore learning is one mode of embedded developmental processes, and in fact is probably better understood as a wide range of consequential interactions with the environment rather than a single or simple thing. How specific it is, both in its topic and how wide or narrow its results can be, are case-by-case phenomena, and much experience or trial-and-error is another variable entirely. Simply put, there is no single "learning" phenomenon to identify.[3]

When learning—in the most familiar sense of exposure, experiment, and study—is involved in development, then the creature absolutely needs certain kinds of experience as raw material, but again, a particular sort with its own particular processing for that species. Therefore the real variable to address would be the parameters of what can be learned. In this, Prendick's species-limitation position is also valid. A pig can be educated startlingly well, even resulting in previously unseen behavior, but it will be an educated pig; one may well say the same about a human. Language capacity is an excellent example; all human infants are already language producers, but they need input and interaction to utilize those mechanisms, and they need the cognitive capacity to internalize and develop it. You won't get a pig to speak merely by altering its larynx. Conversely, that a parrot can say a word has nothing to do with what it may be communicating by it.

To turn a creature's current parameters for cognition into a human profile, including language if you like, then the techniques must target the cortical neural correspondence to perceptual biases of all sorts. Perceptual biases refer to what an animal finds relevant and actively includes in its cognitive processing, both during development and in its later functioning behaviors. Concretely, how would such an altered rat be attracted to creatures like itself rather than to ordinary rats? How would its range of preferred foods, its aseasonal and non-fertile sexual arousal and proceptivity, or its spectrum for aggression be established? Reverse all these concepts for the corresponding human-to-rat switch. Our understanding of these precise neural mechanisms, in their extensive species-specific diversity, remains minimal. That's what would require the groundwork research in this project, at this point a dizzying prospect.

The word "program," the favorite stand-in for "instinct" since computers became household items, is not very helpful. Developmental details for complex behavioral capacities are apparently species-specific, but that doesn't mean these details are not themselves dependent on environmental, interactive conditions, or that they are irreducible units of inheritance and function. They must be addressed from the experiential side as well as the anatomical, because living functions are not a matter of forming an embryo and activating it at birth. Instead, behaviors begin to develop as soon as physiology does. The resolution to Moreau's and Prendick's disagreement applies to the embryo's immediate environmental experience as well, in that you need the circumstances of both a pig's uterus and a pig embryo to get a pig, in so many variables that it's exhausting to imagine summarizing them. For this project, the baseline research to establish the ground rules for the uterine environments and microenvironments is an equally open topic.

As long as we're talking about language, I think it's overrated and should be treated as one variable among many in this difficult category. It's especially debatable whether being able to verbalize things symbolically really matters when it comes to what can be conceived, communicated, and acted upon. The scientific community is still, funnily enough, incoherent about what nonhuman language is even supposed to look like, partly because it's so easy to assume that whatever it is, the nonhumans aren't doing it. I still don't think we know whether chimps or gorillas communicate in language or not. They're so good at solving the researchers' assessment techniques as a social puzzle that they effectively defeat our measures to assess what they're doing.

It's probably inevitable that the project will concentrate on hands and brains, which present a nice example of two kinds of intervention. For all of our talk about the "fantastic and phenomenal" human hand, it's actually not much modified from that of a salamander—thus more ancestral in form than the forepaws of most other mammals. Therefore getting a humanoid hand from a developing rat-like forepaw would require inhibiting its development rather than adding to or "advancing" it, although not very much compared to the far more altered paw of a cat or dog, to say nothing of the single-fingered and hooved endings of some other mammals.

The brain is anatomically more like the stereotype, though, or at least one piece of it is. Most of our brain is boring and ordinary, exactly what you'd expect for a mammal of our size, including, interestingly, the limbic system, which Moreau was understandably having trouble with in the novel—he should have just left it alone. The big size-and-shape difference lies in the outer six layers of the neocortex, present in all mammals. In humans, it is not only freakishly huge—which makes it look like a bloated, wrinkled space-alien—but also contains much denser interconnections among its neurons than are found in most other mammals. Therefore these two features would have to be induced to expand and proliferate, rather than be inhibited.

As a comparison to both of the previous examples, consider metabolic rate, which, compared to body size, is pretty boring and ordinary for both rats and humans, and would not be targeted for a deliberate change. Therefore insofar as metabolic rate influences aging, a long-held concept, perhaps merely scaling up the experimental rat subjects' size will do the trick for human-type longevity, or at least get partway there—except that recent work shows that an array of proteins in cell membranes may be more important in the species diversity of longevity. Unfortunately (logistically speaking), a lot of what look like ordinary traits are going to uncover similar differences in detail that throw a wrench into the project.

Techniques

Again, we're starting from early embryos, and we're going to tweak it so it grows up with the anatomy and physiology we've outlined. How might this be done? By using various means to alter the morphogenetic field at specific locations and at specific times.

The morpho-what? Good question. The term was coined a long time ago, in 1910, but it wasn't appreciated as the powerhouse it is across all of biology until the 1990s, and it took another decade for it to show up in basic biology texts. Hold onto your seat because this is serious business.

You existed first as a single cell, which duplicated its genes, split into two cells, and then each of those did that again, and (insert about a zillion begats) now you are made of trillions of them, almost all clones of that first one. But—if that's all it took to make your body, you would be simply a big blob, with no structural cohesion at all, and therefore a puddle. Soon, any cells surrounded by others would die, as they would have no access to oxygen or any other constantly required resource, like water and sugar. Therefore, in order to develop into an organism, various layers and locations in this rapidly expanding blob need to become segregated in both structure and function, to produce body cavities, systems for bulk transport, a platform for locomotion, a reproductive subsystem, and chemical and neural regulatory systems integrated with all of them.

For any given piece of the body, this occurs in two steps: morphogenesis and differentiation. The first one is the freaky part: when cells that show no individual specificity—that is, they're extremely generic and boring in function—cluster together into recognizable shapes, like the tubes of your circulatory system, or the bean-shapes of your kidneys. All your organ systems are set up this way, in shape, but made of these "blah" cells that effectively do nothing but live, and thus the organs are not yet functional. At this point, each piece we're looking at is effectively made of cellular play-dough and has little or no integrated functions or transport. Since each cell is effectively on its own physiologically speaking, the overall body cannot be very big, especially in terms of volume, or cells would die from lack of oxygen and glucose.

The second step, differentiation, occurs for each organ system at different rates, such that the circulatory system, for instance, differentiates well before the others. This is the part that's easier to understand: various genes in the cells composing each tissue at a given location are activated, and others are inactivated, so that the cells, and correspondingly the tissues, take on very recognizable, very functional shapes. At this point, when you look at the heart, it's no longer cellular play-dough in the shape of a heart, but is composed of muscle cells, all of which are now going lub-dub, lub-dub, in unison. And when you look at the kidneys, totally different genes are doing their thing in those tissues, so the distinctive cell shapes and subunits called nephrons can now be seen, carrying out their various activities. At this point, cells receive oxygen and glucose through fluid transport systems, so the distance limits on diffusion are no longer a constraint, and the embryo can grow rapidly.

Biologists are understandably fascinated by morphogenesis, and Karl Ernst von Baer, who first described these events in the 1820s and 1830s, is considered one of the leading intellects of historical biology, easily rivaling Darwin and Mendel for those in the know. How do these undifferentiated, individually totipotent cells move, proliferate, divide, and group, such that the play-dough version of the creature is established? Because this is how every big multi-celled creature develops, by setting up its parts during morphogenesis, and then differentiation makes them work.[4] The project is merely to adjust these processes experimentally.

The key players in morphogenesis are several (although surprisingly few) proteins called morphogens. If you think of the early embryo as a ball of cells rapidly increasing in size because the cells are continually dividing, then the morphogens operate as local influences on how much the cells divide, how much they stick together, or even whether they live or die in that spot. Morphogenesis proceeds insofar as the various morphogens are released at the right locations for the right amount of time, at the right time. Two things make this more complicated. First, the place in the embryo that releases a given morphogen may not be where it has its greatest effect, because at this stage of development, a protein spreads through the cellular matrix of the embryo like an inkblot across paper, except in three dimensions; this is called a "field." Second, a morphogen's specific effect often depends not on its presence versus absence, but upon exactly how concentrated it is in a given spot.

For example, put your finger at your shoulder joint, press a bit, and then move your finger all the way out to the end of your middle finger. As you go along, you'll find that there are more and more bones involved, both sideways (e.g., two in your forearm as opposed to your upper arm) and lengthwise. This is because during your embryonic development, a single surge of retinoic acid was released at your shoulder, and as your limb lengthened, this morphogen diffused down its length, getting less and less concentrated as it went. The more retinoic acid present when the initial skeletal morphogenesis occurred and cartilage was laid down in it, the more the early versions of the bones fused together. So at the end of your arm is a fan of segmented bones, instead of another big stick like your humerus. So much complexity from one little physical effect, diffusion, of a single substance!

Morphogens typically work like that, practically simpleminded in their immediate release, but when produced in a given spot in a given quantity at a given time, profoundly consequential. Other interesting details include the homeotic genes that affect body segmentation, and the all-important instances of organized cell death, or apoptosis. For example, your hand has fingers instead of a paddle because the cells between the (future) fingers up and died, effectively sculpting the hand before its internal anatomy differentiated. A lot of your body cavities initially formed that way, too.

The real beauty, though, lies in the observation that animal morphogens are apparently closely conserved throughout evolutionary history—in other words, although you and the nearest fruit fly differ drastically in body materials and form, you and it used very similar physiological devices to organize your bodies. This conservation often goes all the way down to the genetic level. If you restrict your attention to the mammals alone, then we are talking about vastly different outcomes from what amounts to minor differences in the use of nearly identical toolkits.

This makes everything so much easier in some ways. It means that a creature's DNA is not a magic blueprint for all its features. Some traits, a very few, can each be traced directly to a single gene, such as the components of our immune system and our blood types, to name one consequential and one trivial example. However, the majority are best thought of as participants in chains of biochemical effects, operating in steps, called "cascades" in the jargon of Evo Devo. Different cascades occur at different times and at different locations in the embryo, influencing the ongoing development.

As a general point, we talk about genes as "information," but we give them too much credit in isolation. Genes by themselves are a handful of idiotic compounds and contain no "direction" or "instruction"—they are one feature of the chemical effects in a context of conditions, interactions, and combinations of events. As Lewis Wolpert (who coined the term "morphogen") put it, they are best understood as *positional* information, phrased by Patrick Bateson as *recipes* rather than blueprints. This phrase has perhaps been diluted through overuse, so I stress that it means the combination of genes involved entirely in the context of the local conditions and available raw materials.[5] In a body, nothing happens, not anatomical or behavioral, except as Gilbert Gottlieb described it for psychobiology, the *entire developmental manifold*.

Fortunately, the remarkable diversity of mammals apparently represents few variations on a limited range of recipes. I like to think of it as the Japanese game called Go: the rules are so simple that you can learn them in a couple of minutes, but a given game can become extremely complex, and different opening moves can influence the whole picture of what eventually turns out to be the winning strategy. Given a few nudges for yea or nay in a few morphogens and conditions at the right places and times, and you could well have developed into a musk ox.

> But perhaps my meaning grows plain now. You begin to see that it is a possible thing to transplant tissue from one part of an animal to another or from one animal to another, to alter its chemical reactions and methods of growth, to modify the articulations of its limbs and indeed to change it in its most intimate structure? (*The Island of Doctor Moreau*, p. 53)

Plain indeed, and at this point, there is no need for tissue transplant or surgery at all. Using human development as a map and schedule for morphogens' release and similar effects like apoptosis, tweaking a nonhuman mammal into a different mammalian species' configuration may well be a matter of finding what morphogens to drip in, or to inhibit, and where and when.

So the watchwords become location, location, location; and conditions, conditions, conditions—as well, of course, as knowing whether a particular outcome-detail gets socked into place early versus what is due to twiddling later. Even if you have the right morphogen in the right amount at the right place, you don't want to introduce it at the wrong time. The point is to mess with the cascade of events in the right way to induce the rest of the cascade, operating normally, to do what you wanted without further need to mess with it. Another procedural parameter is to consider whether you want to change the substance affecting the tissue, or the tissue's response to the substance.

Here are the ways in which a morphogenetic field can be influenced.

- Genes: That is, inserting content into or deleting it from the embryo's genes very early so that all the developing cells will receive this new sort of instruction. It's probably the most laborious way to change the embryo, especially since the effect involved would have little or nothing to do with altering the gene product (protein) and everything to do with its timing. The current knowledge base is pretty slim for these techniques at this level of intervention.
  - Certain single-protein traits, traceable to single genes, may have to be altered this way, especially for the immune system.
- Morphogens: This is what I described earlier—literally reaching into the developing embryo and altering the chemical concentration of a given morphogen (or an equivalent, like growth hormone or an apoptosis inducer), either by adding it or by adding some substance that inhibits it.
- Physical intervention: A somewhat cruder version of the previous method, effectively holding down or forcing apart or otherwise physically intervening with the developing tissues.
- Gene therapy: This is gene modification, too, but only at a given place and at a given time; it might be important to generate a one-time but very significant "bend" in the cascade of a given creature for some crucial bit of anatomy or eventual physiology.
- Conditions: Providing changes in heat, pressure, salinity, or any of dozens of other exterior effects that turn out to be consequential for the development of a given type of creature.
  - Uterine environment issue: Same or different carriers; effects and tolerance regarding multiple in-and-outs—techniques need to be refined for minimal removals and maximum impact.
- Stem cell bank: One for each subject, so cells can be modified externally (genetically or otherwise) and sent into the embryo at key points.

If this is striking you as impossible, well, it pretty much is at present. But to focus on one element, bear in mind how much more we know about the genes compared to even twenty years ago. Both human and rat genes have been fully sequenced, and one of the big game-changers of recent years is called genomics, the technology that permits not only mapping the entire range of genes of an organism, but also measuring their activity. Since the real meat of development lies not in single genes and single proteins, but complex interactions among them, the new technology makes it at least conceivable to measure such interactions as epistasis, heterosis, and other such jargon-named events directly, which is exactly what this project would rely upon for its initial procedural designs.

I know I keep harping on this, but it really bears thought that the project's biggest intellectual barrier isn't the biochemical technique but the developmental, contextual one. Since we know very little about the mechanisms establishing a given species' parameters and biases, all the genetic and morphogenetic manipulations would be like poking our fingers around in a dark room. For this actually to be a scientific project, much more groundwork about the necessary and effective experience of development remains to be laid first. A whole library of techniques and a language for discussing them would probably have to be worked up first from experiments with mice and zebra fish.

For example, human cerebral cortex size and density do not in themselves reflect an increased competence of thought. A lot of nonhuman animals with average-sized and ordinary-density cortices can process things cognitively a lot better than we can; rats beat the pants off us when it comes to spatial memory and the probabilities of which tunnel to try next. If they could read and write, our recreational mazes and Sudoku games would strike them as fit only for toddlers. More subtly, among social animals, the capacity for social interactions, ranking, and manipulations varies widely across species—and at the more complex end, we might find ourselves clueless and baffled among the intricacies and obligations spanning, for instance, a group of baboons or hyenas. If you want to make a functional human out of a chimp, that would entail reducing the capacity for social manipulation, not increasing it. It's quite false to think the expanded, denser human cerebral cortex indicates generalized expanded thinking abilities, or that merely influencing it to be there anatomically would also generate its ordinary function in humans.

Combining Species

Would a version of the project contain such fanciful goals as Bear-Vixens, Mare-Rhinoceri, and Hyena-Swine? It may seem arbitrary in the book—if you want to make a human out of a wolf, why complicate things by throwing bull tissue in there, too? However, combined structural components may be needed, in some cases. To make a cat's forelimbs function in a humanoid way, they're going to need clavicles, for instance, and one way to do that is to graft in early-stage clavicles from some other creature.

I remember seeing my first chimera in 1989, when I was interviewing for graduate studies and happened to attend the departmental seminar being given that day at the University of Maryland. I don't recall who the speaker was, but he was certainly either working with Nicole le Douarin, whose chimera system had been recently published in *Science*, or with a lab directly applying her techniques. I arrived a bit late, so I encountered the slides cold.

My first reaction was, "What the hell is that?" Why was I looking at a plate of goo holding a partly developed chick embryo, a whole side of which displayed recognizably distinct quail anatomy?

Let me explain my reaction. Like all modern biologists, I consider life to be cellular, or, phrased at its most extreme, that only cells are indeed alive. In that context, organisms like you, me, chicks, and quails are cooperating collectives of cloned cells, which take on interesting properties when they're big and complicated enough. All I was seeing was that cells could be induced to multiply and form organs—what they'd be doing anyway—even when stuck to another set of cells already doing the same things. I recognized it as a technical success at a difficult logistic task, but at first glance I didn't see it as answering a recognizable scientific question.

Listening, though, I remembered that cells move around the body a lot during early development, so at the least, the introduced cells make a great marker device for tracking the routes. Later, I would find I'd luckily encountered one of the early stages of the Evo Devo revolution, laying the groundwork for not merely combining two creatures, but affecting the functions of one with the other. Le Douarin–style techniques could be instrumental in getting the developmental template of creature A to tweak in a direction that itself might not be that of creature B exactly, but that yields the human-type experimental goal.

The big problem of rejection remains: a chimera does not survive once its immune system matures, as it rejects the intruder organs with severe effects, and they are typically euthanized before that happens. This becomes much harder when species differences are involved, not due to gross incompatibility, but to subtleties that aren't detectable until they're found the hard way. Stephanie Fae Beauclair (Baby Fae) was kept alive with a baboon heart for twenty-one days, but unfortunately not long enough for a human donor to be found due to a subtle immune incompatibility. To date, using pig organs for human transplants remains hypothetical for the same reason. Some clues to solving that problem may be found in marmoset twin studies; it turns out their cells have lower numbers of receptors to detect intruders.

A more subtle approach might be to get the recipient animal's tissues to respond in a permanent fashion due to transitory contact with chimeric tissue—that is, the introduced tissue would not last or be expected to integrate with the recipient, but it would bring about a given specific response.[6]

But Why

Research doesn't happen in a vacuum. There's no body of devil-may-care, off-the-leash, fully funded and equipped "scientists out there" doing whatever comes into their heads, any more than there's a white-robed, soft-spoken Council of Science vetting every last project for ethics and its purported benefit to humanity. Scientific work today is a plural, variable institutional phenomenon, and most definitely, you don't get to do anything that isn't paid for in some way.

This whole endeavor would therefore either be some kind of Manhattan Project, or it would be scattered around many, many researchers working on bits and pieces and related elements all over the world, probably all arguing and disagreeing over everything, with no set schedule and piecemeal results, which would produce tons of insights but no viable outcomes.

In its engineering phase, however, the project would have to be more central-ized and institutional, for the logistic reason that its subjects must ultimately be gestated and mature normally. There is no such thing as a people jar, such as you see all the time in science fiction cinema and TV, nor is there any viable means of accelerated gestation and maturation. For the rat-ending subjects, that means born with rats, nursed by rats, reared by rats, and living in rat environments, for this purpose let's say an extensive semi-natural one skewed heavily toward rat preferences, living quarters, and patrol spaces. For the human-ending subjects, if we're looking for human-style development and performance, that will require human-type levels of gestation, nurturing, learning, and maturation time, in a functioning social environment. And that means a genuine human environment, ordinary families in ordinary society, as a serious design parameter. You can't keep them in a stereotyped science fiction "facility," because if you marginalize them, you interfere with the performance variables. Don't forget the double-blinding, such that none of the subjects, the families, or the people recording data would know which individuals came from which original species. Arguably, the subjects and families wouldn't even know about that aspect of the experiment at all. The final "do it" step would itself be huge, to bring up all those kids, so it would need serious infrastructure.

I raise this to strike to the heart of the Man/Beast false dichotomy. In the novel, Prendick first mistakes the Beast Folk for experimentally mutilated humans, which drives no less than seven full chapters of the story. His error isn't irrelevant, but instead fits right into the problem at hand. Specifically, why does learning that the direction of Moreau's work is reversed make it less wrong to him?

Let's say the techniques are eventually rock-solid, amazing, and we get lovely rats from both the experimental subjects and their positive controls. Some of them were born from the pairing of human beings. Consider them carefully, the rat pups in this project, regardless of parentage. They live in their rodent environ-ment, sleeping, patrolling, eating, and socializing in rat fashion. They squabble, huddle, burrow, and yes, mate like rats. Even if the experimental subjects happen to be non-reproductive, that doesn't stop them from giving it a good try. They last for the several years that rats get, they age, and they die.

Similarly consider the kids in this project, both the other group of experimen-tal subjects and their positive controls. Some of them were born from the pairing of rats. They are human children: raised in families, probably observing various religious practices, going to day care, going to school, wearing backpacks, playing with favorite toys, finding friends, and making a place in their families and social lives. They get sick and they get hurt, and they need family help and support. They learn, laugh, and cry. They're loved and included just as any children are, which is to say, to varying degrees. They grow up, they have favorite music, squabble in their pre-adolescent and adolescent cliques, and they seek an adult identity.

If you're horrified, is that because the project would treat humans as experimental subjects like the rest? Is it time to talk about the dignity of humans who by every imaginable metric are experiencing the absolute minimum of stress and pain? Or because it would treat a member of another species exactly like a fellow human? Is it time to ask the same question raised regarding human-on-human bigotry: "Would you let one marry your sister?"

As if that weren't enough, consider this: because the variables in this project are all so graded and scheduled in terms of small steps, a given series of subjects should be terminated and assessed just to see whether the earlier steps are working. So that's a huge number of embryos that would be started and stopped almost right away; then if those seem to have worked well enough, then a next wave that would be taken a step or two beyond that, and so on. Such a step would probably have to be repeated later, too, if a given phase of the development turned out to be balky and to require isolated testing, or even new such steps might have to be imposed. Since so much of development is postnatal, extending into adolescence, that raises the issue of terminating individuals: Should that be done to the rat-origin human-like kids who are already well into their childhood? If instead the subjects were removed from the project, tended to as best we might, and permitted to age and die without intervention, that's pain category E in the IACUC schedule—for animals, the most extreme category, the least permitted relative to the other features of the project. Why is euthanasia considered the most ethical end for a lab animal but the least for a person?

Or take the issue of their legal status relative to being under scientific observation. It's not *ipso facto* illegal or wrong to study humans scientifically. Humans are the most extensively studied species on the planet; we have voluntarily normative observation, voluntary clinical trials, and use of already available information across non-planned groupings. In a study of this kind, with no pain and no limitation of movement, with only normative observations, permission would lie with parental consent, or with those empowered to act as parents. Is that what applies here, too? I assume that few or no parents would permit their children to be in the rat-ended experimental group, but would it apply to all the children with human features, human origin or not? In fact, are the people who raise them—who ideally would not know their origins—legally their adopted parents? Is this, you know, *real* adoption?

At Last

At last, we are talking about research ethics that have nothing to do with Bernard, nothing to do with the two windows—never mind either "agony for knowledge" or "the higher purpose." The issue lies in our inability to process that humans are a kind of animal. This is the freak-out moment I described in Chapter 3, because the leveling of human and nonhuman already exists, but is now made visible and put into practice. It necessarily places humans of nonhuman origin into our own society, and therefore this project grades right outward from development, physiology, and evolution into social engineering. Individual compliance with professional regulations is no longer the concern, because it's not single-study ethics at stake. Instead, the challenges are made to and demands made upon policy, which means—just as Biller-Andorno has written—this concerns justice. One cannot blind oneself to ethics in order to work with live animals, as Bernard says and as I think is still implied in the current activist discourse, but rather should embrace this as an ethical task, fully articulated and in a community effort.

So, what is justice, regarding nonhumans? At all? Within scientific use? Certainly my injured vole deserved some, and given the changes since that quarter-century ago, I think that current dialogues and use-practices in science have considerably more justice-oriented content than they receive credit for. But the concern of such practices is still focused upon pain and suffering, and complex as that's been, it's comparatively easy. Justice in regard to this project is another thing entirely.

Current scientific standards are not sufficient to handle that because nothing in our cultural vocabulary, for this issue, is sufficient to handle that.

For example, does injustice lie simply in being a human being? Is it cruel to be made into one? If not, then is it wrong to alter a nonhuman into a human? If so, then why? Because it is unworthy of presumed elevation? Is it wrong to treat these individuals as human beings (or rats), or wrong not to? How are they treated: people or not, rats or not, treated well or not? Are they citizens or property?

The rats in the project aren't proxies, which is to say, stand-ins for human subjects due to certain common parameters or for useful comparisons. There aren't any stand-ins in this study; each species is there as a comparative point for the other, on equal footing in terms of biological interest. Since the techniques by definition equate their biological status, some case can be made for all of them to be treated as people, legally—as human subjects. And then we have to debate whether that applies to all four groups, regardless of outcome, such that a completely ordinary rat resulting from the control group would legally be a person for the rest of its life.

So far, I've been discussing *animal welfare*, but the term "animal rights" is established as a different thing—the concept that the use of nonhuman animals is an inherently unjust thing, subject to criticism on the same basis of the denial of human rights. The concept was introduced by Tom Regan in his *A Case for Animal Rights* and has gained considerable traction.

The argument is strong in many ways, not least in that it rightly calls out the classic naturalistic fallacy inherent in the concept of human rights insofar as they rely on a special status for humans. Taking as a given that people, all people, have rights regardless of personal abilities or specific performance indicators is a fine assumption as far as I and most people are concerned, until one realizes that its rock-bottom justification is merely that we belong to a particular species. The critical term for that is "speciesism," coined by Richard Ryder, which is the assignment of literal value and consequential legal protection solely on the basis of belonging to a given species, and as such, patently a matter of maintaining privilege. Citing "humanity" is arguably no more than exploiting a loaded variable that is devoid of moral or other higher/lower content, and it should be more sensible to refer to the less species-specific concept of dignity. Animal rights advocates call for a reassessment of nonhuman animals' legal status, in many cases supporting the position that current scientific animal use is a rights (dignity) violation.

Let's take a look at that, because I admit to some confusion. So far, all discussion of rights begins with the assertion that humans have them. It's essential—and I choose that word precisely—that Man, in the Man/Beast divide, is gifted not only with special abilities, but with a special status, a way to be treated. Much discussion of rights concerns how we can get them to people who are denied them, wrongly denied because they too are Man, and the discussion of animal rights simply extends that same question to more creatures. This strikes me as an arguable position, although some of the justifications baffle me. I have not been able to grasp why many of the arguments analogize nonhumans with disabled people because they cannot talk, possibly because I think speech is an animal ability, like a bird's capacity to fly, so I can take another animal's dignity as given without the need to cut it slack.

A deeper look exposes human exceptionalism, though. This claim relies on the existence of the special status as a profound and mystic thing. Rights in the abstract are an example of religious thought. To say that rights come from God is more honest than to say we "just" have them in some ineffable and unquestionable way within an otherwise material existence. God may have gone out of fashion, but the "rights of man" still rely on Man as a special status, as metaphysical as ever. Adopting some other species into that category maintains the category, unless all life on earth is included, which seems a bit odd since rights are supposed to be a method of arriving at policy. Put most harshly, nobody actually has rights or dignity—as soon as you leave the realm of abstraction, you're back to the grosser realm of what you're going to do.

That brings up the practical problem with ineffables, that having them on paper, or in the ether, as it were, guarantees nothing about having them in a tangible sense. In practice, rights need to be named and applied. In this case, are we talking about protection from one or more forms of harm? Being permitted a wider (free) scope of action? And for whom?

The application is the toughest, because the designation of rights is not a force field. You may have the right to live, but that by itself doesn't mean someone isn't going to kill you. Here I distinguish between *talk* of rights and *actions* of humans toward one another, which includes all too much slaughter, misery, and exploitation, and not just from "anti-social" or "barbaric" people either, but mainly through institutions of vast organization, wealth, and power—the civilized ones, I believe they are called. When I look at the action relative to the purported rights, much of it concerns how we can deny them to people whom we happen not to like, or from whom, bluntly, we gain advantage by killing or immiserating. This turns out to be easy because the perpetrators can simply say they are not Man. Unfortunately, in practice, rights are circular: we *say* we have them because we are Man, and we designate who is Man by assigning them rights. This is not about inherent and essential qualities, but about inclusion and exclusion.

Perhaps Man, out there, up there, somewhere, not here, does indeed have rights. Good for Man. But you and I and everyone we know aren't him. Moreau would be disgusted by us, just as Prendick is in Chapter 22. We're grubby, confused, ordinary, committed to one practice or another, trying to talk to one another about what to do next, able to cooperate and able to fight. With much pomp and verbiage, we include and exclude other humans in our spheres of power. Everyone calls the way they do it, or want to do it, "my right."

That's why I can never quite get the charge of speciesism, or at least, what seems to be rock-solid certainty that assigning nonhumans rights would turn into genuinely better practices toward them. In the constitution of the nation where I was born and am a citizen, slavery is barred—except for convicts and conscripts, as stated in the Thirteenth Amendment to the US Constitution. At the time of this writing, this nation leads the world in its absolute number of imprisoned convicts, in the length of incarceration, and in the practice of solitary confinement. Convict labor is leased to private industry in an explicit act of chattel sales. The convicts' status as slaves, badly treated ones at that, is blatantly apparent in any terms you care to name. I guess their rights don't matter? Or they don't have them? Which is it?[7]

Is that what animal rights activists want for nonhumans, the same discrepancy between the paper rights and the power-based, provisional reality that real-life humans experience? In that case, I don't see the difference that assigning rights would make. Cows would get well-phrased, resolved, and ratified rights, except, you know, for the cows we eat. If we can't get human rights universally applied to humans, then I don't even know what we're talking about. I find more value in the now thirty-year-old practice of applying the Animal Welfare Act to lab rodents, despite their not being covered by it.[7]

This isn't about them, it's about us, because we have the power and the capability. The question is what we will or won't do, letting "why" remain the community quilt that it is, rather than an arrived-at truth. In that case then, in this imaginary Moreau project, are the rats' rights being violated? In fact, let's go all the way and legally place the experimental subjects of rat origin into the status of humans as raised by humans, subject not to the Animal Welfare Act and associated regulations, but to the provisions of the Nuremberg Code and the Belmont Report. If that were so, is this project a violation of their nonhuman rights? Is a given creature's dignity violated, although it now has exactly the dignity of another kind of creature? Would treating the rat-derived people as rats rather than people be appropriate due to their "native" origins?

The Moreau experiment doesn't "challenge natural law." It's flatly confirmatory of current thought on evolutionary history and humans' biological identity, as there is simply no technical controversy in the phrase "human beings are a species of animal," or in "species' differences result from changes in developmental mechanisms." But just as SCINT cloning disrupts nothing but the *familiar timing* of ordinary twins, and thus prompts a social and legal freak-out regarding what "family" means, a Moreau project disrupts nothing but the *familiar designation* of ordinary species, and would clearly prompt an even more profound freak-out regarding what "human" means, including issues of human rights, human interests, human dignity, or anything else tagged with the term.

Fortunately, we have a book to talk about instead of being confronted with it cold. This is the brilliance of *The Island of Doctor Moreau*, because it disrupts the entire ethical and political discourse, whatever degree of animal welfare is under negotiation, whatever rights of humans or nonhumans are adjudged to exist, and whatever side of a currently designated controversy one may fervently support. Pain is only the opening topic, and even when that issue is removed, the novel still challenges the sides' very existence to expose the exceptionalism within.

Readings

The literature of Evo Devo is among the most exciting scientific reading available, including Theodore Garland and Michael Rose, *Experimental Evolution* (2009); Alessandro Minelli, *Forms of Becoming* (2009); Mark S. Blumberg, *Freaks of Nature* (2010); Lewis Held Jr., *Quirks of Anatomy* (2009); Wallace Arthur, *Biased Embryos and Evolution* (2004); Mary Jane West Eberhard, *Developmental Plasticity and Evolution* (2003); Benedikt Hallgrimsson and Brian K. Hall, *Epigenetics* (2011); Rudolf A. Raff and Thomas C. Kauffman, *Embryos, Genes, and Evolution* (1983); Günter Wagner, *Homology, Genes, and Evolutionary Innovations* (2014). Great credit is also due to Stephen Jay Gould for the early contribution, *Ontogeny and Phylogeny* (1977).

References for human subject controversies include the Tuskegee syphilis studies and the Nuremberg Code. The primary current reference in the United States is the Belmont Report, published in 1978–1979 by the National Commission for the Protection of Human Subjects of Biomedical and Behavioral Research. Some provocative examples are found in Rebecca Skloot, *The Immortal Life of Henrietta Lacks* (2011); and John D. Marks, *The Search for the "Manchurian Candidate"* (1991).

Primary texts regarding technical rights for nonhumans include Tom Regan, *A Case for Animal Rights* (2004); and Richard Ryder, *Victims of Science* (1975) and *Animal Revolution* (1989). For summary and further debate with a strong grounding in biology, see James Rachels, *Created from Animals* (1999); and Cass R. Sunstein and Martha C. Nussbaum, *Animal Rights: Current Debates and New Directions* (2005). Matthew Scully, *Dominion* (2003), develops the issues of agency and justice for scientists, rather then seeking intrinsic qualities within the subjects.

Notes

1. Eric Olson, "The Developmental Renaissance in Adaptationism," *Trends in Ecology and Evolution* 27(5): 278–287, 2012, published by Elsevier Ltd.
2. Sherryl Vint's points about anthropomorphism and theriomorphism in *Animal Alterity* (2013) are relevant here—we objectify nonhuman species' features and status precisely as we do our own.
3. This construction of learning also applies to purely anatomical variables—in order for the limb bones of vertebrates to develop, they have to be utilized as the creature learns to walk. Genes don't simply gift the creature with functional legs; instead, the process of learning to walk, the interaction of neurology, biomechanics, and gravity, helps to build the materials with which it's done.
4. Cells' mechanisms for recognition and signaling weren't identified until the 1960s. Until then, developmental processes did seem almost supernaturally directed, and it's no surprise that developmental mechanics had always been the sticking point concerning evolutionary processes. Once the cellular mechanisms had been exposed, however, both theory and experiments about development boomed.
5. With the term "recipe," I am not referring to a single gene's activity in protein synthesis, but rather to the activity of many genes at particular intensities and at different times, in the presence of specific raw materials and conditions, which results in a functioning physiological structure.
6. Phillip Karpowicz et al., "Developing Human-Nonhuman Chimeras in Human Stem Cell Research: Ethical Issues and Boundaries," *Kennedy Institute of Ethics Journal* 55(2): 107–134, 2005, critiques the idea of human-nonhuman chimeras as an affront to human dignity at several levels. To my eyes, their reference to *The Island of Doctor Moreau* raises the interesting question of whether Moreau's experiments would in fact "degrade human dignity" because no original human cells or tissue were employed.
7. It may clarify my point to explain my position regarding the Civil Rights Act of 1964, based on scholarship which holds that the Act itself was not instrumental in securing better circumstances for black Americans, and that in the areas most associated with the Ku Klux Klan and other murderous and humiliating practices, more credit is due to the Deacons of Defense and other armed security and resistance groups, at the very least in combination with the peace marches, boycotts, and the Act. The argument is that legislation is not the single most valuable solution or end-goal, but is at most a useful part of a multipronged, even decentralized, set of efforts. Useful texts include Lance Hill, *The Deacons for Defense* (2006) and Charles E. Cobb Jr., *This Nonviolent Stuff'll Get You Killed* (2014).

PART III

POOR BRUTES

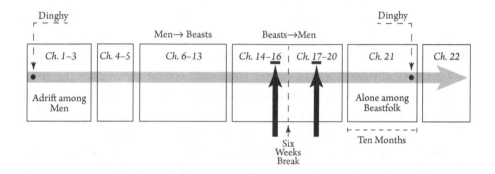

> In fact, the conviction that the world and man is something that had
> better not have been, is of a kind to fill us with indulgence towards one
> another. From this point of view, we might well consider the appropri-
> ate form of address between man and man ought to be, not *monsieur,
> sir, mein Herr,* but fellow sufferer, *Soci malorum, compagnon de misères!*
> —Arthur Schopenhauer, "On the Sufferings of the World,"
> *Studies in Pessimism,*[1] p. 216

Human. Nature. Human nature. Let's see how people talk about it.

You left your wallet on the photocopier machine, realized it while going
out to your car, and returned to find it gone. "Human nature" strikes again,
right? But then, as you're back in the parking lot, on your phone and com-
plaining to your friend, who's saying those very words, an unknown person comes up with your wallet and says, "Hey, I just found this, I've been

looking for you." All your things and money are still in it. Was that "human nature," too? I bet your friend doesn't use the term then. No, that wasn't nature. It was cultural. It was their upbringing. It was their goodness. It was—wait for it—an act of our "fellow man," or "simple humanity."

OK, wait. "Human nature" and "simple humanity" are two different things? Humanity is wonderful, but human nature is the barely submerged beast? The admirable features are intrinsic to our species, even "simply" so, but they're ineffable, whether you want to call it spiritual, psychological, rational, cultural, or moral. By contrast, the phrase "human nature" most often refers, strangely, to something bad, or at least not as good as these things, shamefully present due to intrinsic fault or to leftover animalism.

There's no point in trying to parse and justify the distinction. It's pure compartmentalization, using code-phrases. The term "human" (humane, humanity, as in "appeal to humanity") designates a special status: what we like, what we consider separated from the Beast, and what we'd like to say we intrinsically are. Human nature is about some bad thing lurking in it. The common addition of "admitting" to some remaining Beastliness is still a dodge, preserving the divide—we must go further.

This part of the book concerns looking at both nonhuman and human animals to see this construct evaporate, and then coping or failing to cope with what one sees. The whole story as narrated by Prendick is best understood as precisely this act. As the old joke about research on chimps says, the first shock comes when one sees that the gaze travels in two directions.

Note

1. Arthur Schopenhauer, "On the Sufferings of the World," *Studies in Pessimism*, reprinted in *Collected Essays of Arthur Schopenhauer*, published by Wilder Publications LLC, 2008.

6

All the Difference

Prendick may be the narrator and the main character in terms of presentation and point of view, but there are many protagonists in *The Island of Doctor Moreau*, for which the human characters (i.e., *Homo sapiens*) are foils. Look at the story again, this time with different points of view creating different through lines, and you'll see complete personal arcs, full of insights, decisions, and committed actions, for four of the Beast People.

- The Leopard Man, in Chapters 9 through 16 of the novel, relative to Prendick, discussed in this chapter
- The Puma Woman, in Chapters 3 through 17, relative to Moreau, discussed in this chapter

- M'Ling, in Chapters 3 through 18, relative to Montgomery, discussed in Chapters 7 and 9 of this volume
- The Hyena-Swine, in Chapters 16 through 21, relative to Prendick, discussed in Chapters 8 and 9 of this volume.

Processing these personal stories is complex: they're filtered through Prendick's narration, and his ideas change several times, always in relation to Moreau's unchanging view. To get into these stories, these more explicit perspectives need to be explained, prior to parting them like curtains.

Moreau's Man

Moreau's point of view is straightforward, unswerving, and incorrect in one crucial point, which serves well as the baseline view of the Man/Beast divide against which Prendick's changes can be compared.

Again, cast aside the "don't meddle" story, which has it all wrong. Moreau isn't trying to *be* God—he is appreciating God by investigating and extending his works, the perfect nineteenth-century scientist with one foot in the Romantics. He reveres "man" as the primary work of God, and therefore the obvious and imperative aim of such investigation. Moreau's problem is that his vision of such a work is already idealized entirely out of the ballpark of being human.

In fact, I'll name his goal, not the ordinary descriptive label "man," but the exalted, exceptional status of "Man." This construct is perfectly rational, spiritually elevated, beautiful to behold, master of the world, emotionally serene, and civil in every way, Spencer's ideal to the letter. Moreau's uncritical and arrogant view of this entity not only permits his own inadvertent sadism, but blinds him to the results before his eyes: that his creations suffer, believe, sin, and struggle to the same extent that real people do. Moreau's consuming passion is to overcome precisely this outcome.

For all the clarity and consistency in the rest of his argument in Chapter 14, he is oblivious to his single, outrageous self-contradiction when Prendick asks him why he chose the human form as an experimental goal. His slick "by chance" answer is completely incompatible with every other description by him of that particular goal, throughout the rest of that conversation as well as elsewhere in the story. He really wants to make, not only a person, not only a human, but Man out of entirely nonhuman material. He's not even trying to create a new kind of "perfect man," because to him, Man as is, the Creator's work, is already perfect.

Making a real one would be enough—but it must be according to the exceptionalist expectation, not "animal" at all:

> ... just after I make them, they seem to be indisputable human beings. It's
> afterwards as I observe them that the persuasion fades. [...]
>   [T]his time I will burn out all the animal, this time I will make a ratio-
> nal creature of my own. (*The Island of Doctor Moreau*, p. 59)

This is the core of the plot: Moreau succeeds with his procedures. But he cannot possibly see this, because this Man, as he conceives it and which he tries to
create, does not exist. He does not see what he has achieved, because he never
sees anything but the absence of the rational man. He throws his subjects out
into the night, saying, "Bah, failure," and has little further to do with them. But
what they do out there, on their own, is to act like and indeed to experience life
as people do.

He complains about them bitterly:

> The intelligence is often oddly low, with unaccountable blank ends,
> unexpected gaps. And least satisfactory of all is something that I cannot
> touch, somewhere—I cannot determine where—in the seat of emotions.
> Cravings, instincts, desires that harm humanity, a strange hidden reser-
> voir to burst suddenly and inundate the whole of the being with anger,
> hate, or fear. (Ibid., p. 58)

Once he gets going on his obsession, his pretended indifference to seeking the
human form sloughs away:

> I can see through it all, see into their very souls, and see there nothing
> but the souls of beasts, beasts that perish—anger and the lusts to live
> and gratify themselves. [...] There is a kind of upward striving in them,
> part vanity, part waste sexual emotion, part waste curiosity. It only
> mocks me.... (Ibid., p. 59)

Aw, poor Moreau. He keeps trying to get a Man, but (damn it!) keeps winding up
with boring, grubby, feeling, variably articulate, unfortunate *people*.

I ask that you reflect upon this for a moment. Moreau is a maniac in the technical sense, rather than the Hollywood sense, arguably psychopathic in that same
technical sense, and all too willing to inflict agony—because of his unshakeable
mainstream perspective. He's crazy not because of his scientific techniques, but
because he *agrees* with the Man/Beast divide—his one central and bonkers notion
perfectly represents what most people think!

The films consistently and accurately define the Moreau character as committed to the Man/Beast divide, rather than questioning it. This is probably best conveyed, if unsubtly, in the 1977 *The Island of Doctor Moreau*, when Moreau flips his lid and starts whipping the poor Bear Man for evincing even a bit of his origins, as he's clearly unhinged at the sight of any combination across the divide.

However, modifying his view so that he seeks to improve upon humanity through his work, as he states outright in several of the films, is a dodge to keep the topic of real people out of the discussion. The line in the 1996 film that claims his creations are " . . . a divine creature that is pure, harmonious, absolutely incapable of any malice" is difficult to interpret because it sits on the boundary between conceiving of real people as perfect rational beings, and seeking to create such a being because real people are not.

The films also struggle with how this essential quirk of Moreau's relates to religion, or more accurately, to God. The common nineteenth-century view that the Creator made a world that operates by laws accessible to science fell out of fashion sometime in the early twentieth century and is not well articulated or commonly recognized today. Therefore the films—which draw upon the strong scientist-as-blasphemer template anyway—either make Moreau a hyper-intellectual whose science defies God and nature, or effectively a cult leader, whether cynical or naïve, whose science is merely a means of creating a subject population toward that end.

If Moreau seems thickheaded about his notions of Man, he is at least in good company, which is to say, nearly everyone else, both in his time and ours. His unchanging view perfectly expresses contemporary Angel/Ape thinking.

These juxtaposed terms "Angel" and "Ape" refer to specific Victorian values and view of humanity, the idea that the human experience is a battleground of two absolutely opposed influences. The first is a veritable army of violent, irrational urges that arise easily and swiftly, bypassing all thought. Sex! Anger! Bad manners! Think of a host of evil spirits or an inherent legacy of sin that infects the mind, then give it a scientific face and gloss by recasting it as vestigial traces of the wicked brutes from which we, humanity, have barely emerged. Meanwhile, the opposition engages in a holding action by progressive, productive endeavors, distilling a perfect social order from human history, all to be instilled in an obstreperous populace only with great effort of some kind (military force, parenting, civil education, take your pick). In this construction, civilized culture is completely and uniquely an add-on to the human animal, rendering it *not* an animal—bonus points for tagging your own culture as especially accomplished in this regard.

Conceiving the human mind, or "psyche" to use the contemporary term, as such a battleground feels good. It is exactly the same exceptionalism in both social-reformer fervor and ostensibly secular (but not really) military-imperial enthusiasm, with religious language added for either, or not—frankly, that particular language doesn't matter because the construct and the ideal are the same. My eyes fall upon a different, less trivial language:

- Beastly, bestial, savage, uncivilized, primitive
  - Sometimes, emotion or passion, when constructed as uncontrollable
- Human, humane, humanitarian, civil, advanced
  - Sometimes, reason, especially when linguistically elevated to the term "rational," allowing for passion only when it's directed toward a worthy goal or cherry-picked to be selfless or radiant in some way.

The same things get renamed in various contexts: nature versus nurture, nature/biology versus culture, deterministic versus choice, genes versus environment, instinct versus learning—always opposed, always the same Manichean battleground. Even switching sides to romanticize the Ape as spontaneous, natural, sincere, or authentic doesn't change the overall construct.[1]

I don't advise questioning this construct outside an organized discussion, based on long experience. The nineteenth-century obsession with *model of humanity* as synonymous with *policy position* is still very much with us, and the Angel/Ape model is the bed from which most of our modern political views have sprung—to question it is too easily perceived as a direct attack upon one's own views, or as sneaky support for the worst view the person can imagine.

I can hear it now: "But, but we are smart!" This is one of the arrows in the exceptionalist quiver: an appeal to functionality as an obvious hard-line separation between this constructed Man and Beast. It's even exactly what Huxley told his audiences in his 1860 "Lecture to the Working Man," specifically referencing speech as the key variable between the two, thirty-three years before he'd present the rather different view in *Evolution and Ethics*. I have bad news for you, though: no, we aren't. We're pretty good at some neat cognitive actions, but it doesn't make us smart in any qualitative sense, nor are those actions all that different from what most or all other creatures already do. And we're really bad at a lot of mental stuff that this or that other creature can do well.

There is no body-mind divide in biology; the concept carries no weight whatsoever. This is right out of Lawrence's lectures, that thinking is an operation, a physiological act that we exhibit. Like any other physiological undertaking, thinking has parameters of competence; and even within those parameters, it may work or not work, it may produce or not produce, and it may succeed or not succeed, relative to any criterion you like.

The terms "sentience" and "sapience" are not observed in the vocabulary of any scientific discipline. Biologists, especially, do not use them because they apply to every organism you ever heard of. All living things continually *perceive* their environments and *adjust* their behaviors in response. These perceptions and adjustments all have parameters and extents according to the species in question, but the underlying point is that there are no such things as an insensible or a moronically robotic living creature.

For instance, do plants think? The simple answer is yes. More helpful is to consider what "thinking" means: whether it's the means by which we behave, in our case, nerves and brains, or the outcome of behaving regardless of particular physiology, which is to say, responsively and strategically. In outcome terms, plants definitely think, but since it's practically impossible to use that word without the vertebrate-centric brain-oriented implication or the human-centric verbally oriented implication, I have to qualify it—with a full dose each of Captain Obvious and So What—by adding that plants' experience of their own behavior is certainly different from ours.

To focus on those outcomes, the real question is, what do real creatures think about? In that context, behavior comes in effectively two forms, staying alive (maintenance) and making new creatures (reproduction). Both can be quite indirect and sophisticated, with lots of subsidiary and derived properties, even including reversals, but details and add-ons aside, a living creature's life transfers stored energy into radiant heat, and when you look at how, maintenance or reproduction is typically what's happening. Here's the thing, though: since a packet of stored energy is used and lost only once, the creature has to trade off its limited stores between the two general categories, as well as among many options within each, including options like "no" and "not this time" in many cases. Life for every creature, no matter how simple, is a nonstop series of literal decisions among viable options. Yes, bacteria too. Everybody.

That line of thought is scientifically powerful, but it is so abstract and divorced from our personal experience—thoughts, identity, emotions, motivations—that it's hard to apply, and the discussion typically shifts quickly from "what about" to "how." In that case, we have to bring it all closer to home and talk about animals, with nerves and in many cases brains. Here's the question of whether we are functionally special, which in fact was investigated startlingly well, practically in isolation and without really good follow-up for a century, by none other than Charles Darwin. It might surprise you to learn that evolution was not really his primary topic of study, but rather nonhuman reasoning and emotions. He might qualify as the single unequivocal non-exceptionalist thinker of that era. However, this work was not well appreciated during his life, nor for about a century afterward.

Past the middle of the twentieth century, the study of behavior remained outside biology as such, conducted primarily in branches of psychology, such as the comparative school founded by Frank Beach and its splinter school of psychonomics, the behaviorist school made famous by B. F. Skinner, the ethological school best known in the work of Konrad Lorenz, and the too-often missed developmental school of Theodore Schneirla and Daniel Lehrman. Other branches include the famous anthropological work with gorillas by Jane Goodall and Diane Fossey. A lot of this work became overrated and a lot of it was unfairly passed over. The disciplinary boundaries and inter-school barriers were very strong, so that most of it remained isolated from formal biology, and much of it did not even get cross-referenced from subdiscipline to subdiscipline. It took a long time for experimental research to focus on problems that made sense to the organisms. Studies of rat behavior didn't start using semi-natural enclosures until the 1970s, for instance.

Academic integration and a certain intellectual housecleaning emerged by about 1980, including the new designations of behavioral ecology, cognitive psychology, psychobiology, and behavioral neuroscience, as well as refining and effectively resolving the long-standing debate about behaviorism. Behavioral work began to connect with biological and evolutionary researchers, at first in informal academic crossover, then in interdisciplinary groups, and finally, today, many scientists include it in biology along with physiology, systematics, ecology, genetics, and cell studies.

It's rude but accurate (and funny) to criticize the historical study of mental operations as a bit schizophrenic, with, as the quip goes, the mind studies being brainless and the brain studies being mindless.[2] However, a lot of the classical work does stand up, including Darwin's, and the modern integration is very, very good. One thing is blatantly clear: there is no Beast process or part of the brain-and-mind that is opposed to some Man process or part, not anatomically, not physiologically, and not behaviorally.

Many brain-and-mind operations are quantitative and organizational, or in plain language, based on counting and comparisons. How much quantity, how much distance, how much time, how much difference, how many repetitions, in what combinations—creatures are all really good at processing this kind of information and basing decisions on it, although quite strictly within specific limits, or if you will, priorities. In the animal, neurological context, this precise kind of perception and processing is called cognition.

Studies of cognition are unequivocal: it's long past time to drop the model of humans with mystic capacities to do ordinary things and of nonhumans as insensate automata, switching back and forth between these portraits whenever considering real observations about the other.

Here's a key point: most behaviors tagged as reflexive are no such thing. The physiological response called a reflex has been quite wrongly applied to behaviors that are simply fast and nonverbal, but rely on as much cerebral processing as the most complicated thing you can think of. Just because you did something "without thinking" doesn't mean it was a technical reflex, as observed in the compromised nerve of your knee. It was thinking all right—it simply wasn't *talking*, and we tend to ascribe too much power to that activity. This same error has also been confounded with "instinctive," unfortunately, with a whole raft of further associations folded in, so that humans' behavior can be falsely walled away from instinct, and nonhumans' behavior can be falsely tagged as unthinking no matter how complex. As I see it, the discipline best labeled "behavior" is close to tossing out all reflexive/processed, instinct/learned, and involuntary/voluntary distinctions, as in their current general constructions, they reinforce the false image of nonhuman living creatures as mindless robots and of behaviors as snap-in modules.

Also, again to varying degrees per creature, perceptions are stored and constantly cross-referenced with both new perceptions and with other stored perceptions from the past, for which "memory" is an insufficient word. This is a significant, consequential operation, but it is neither independent of brain function nor limited to our own species. One of its effects is literally an environmental context for many details of behavioral development. Another is to generate an inner, imagined world, or "constructed" if you prefer, as vivid and important to the creature as the outer, real one, and—from the inside—not easily distinguished from it. Biologists and experimental psychologists call this the cognitive map.

We have no idea what other species' literal experience of their cognitive maps are like, whether they "imagine" in the same way we do. For ourselves, I stress that the word "imagine" is a bit misleading because I'm talking about more than that—I'm talking about when you walk into a lamppost or other stationary object on a street you're perfectly familiar with. It's because you weren't consulting your immediate sense analysis, you were consulting your cognitive map.

I'll focus on one well-defined social subset of perception and cognition: communication, in which an individual perceives the signals of another and acknowledges this receipt to the sender. (Its definition is quite careful to distinguish communication from, for example, merely noticing what another creature is doing.) In a highly social species, individuals constantly "swim" in communicated information from other members of the group and live in that "sea" as much as they do in the plain-and-simple physical world. Our own combination of immediate input, memory in this broad sense, and communicated content composes our experience of life, and may be thought of as the medium for our behavior. The position of oneself as an individual in this map is a completely functioning variable, too. From inside, we dress it up as "consciousness," "awareness," "imagination," and "being smart"—and in pure exceptionalist glory, claim it for ourselves all the way from the ground up as a unique feature. I have no beef with the idea that our exact version of it is species-specific, but denying similar, specific versions for other social species is simply to deny the entire body of evidence. Creatures cognitively compare multiple possibilities, assess immediate conditions in the context of past experiences, imagine multiple outcomes, and judge distances in time, amounts, space, and social relationships. Call this "being aware" if you want, but don't pretend it's a unique human capacity.

When you put the brain into the study of the mind, and the mind into the study of the brain, remarkable work can be done. Here are just a few of the studies from the past few years, taken from a high-powered series of biology journals called *Trends*:[3]

- Regions of the cerebral cortex
  - Two-dimensional, or area-based zones of activity set limits on attention, recognition, and memory, but in the long run, the borders shift in what appears to be a competitive manner.
  - Two different sorts of self-activity are identifiable, an active agent ("I) and a reflective or perhaps narrative product ("me").
- Cell functions within the cortex
  - Neural impulses among specialized pyramidal cells travel outward through the layers of the cortex and then inward again, regulated by tiny, local inhibitory processes called microcircuits.
  - A particular receptor type (kainate-type glutamate) has diverse and complex effects at the juncture of cortical cells, regulating their excitability.
- Remodeling and plasticity
  - Microglial cells eliminate and generate the synaptic connections among cerebral cortex cells all the time, such that the neural circuits are constantly changing (plasticity).
  - This remodeling goes on all the time in both the cortex and the hippocampus, in response to particular kinds of stimulation, as a feature of ordinary behavior.
- The hippocampus is the site for temporal processing: memory, time and spatial orientation, and prediction.
  - Whole neuronal ensembles manage how we remember order and sequences, but single time cells and place cells encode the moments and locations.
  - Forgetting things arises from a built-in decay process (efficient pattern separation), but difficulties with encoding and retrieving memories in the moment result from interference—in other words, "forgetting" is a lot of different things.
- And taking comparative work to its arguable and fascinating extreme
  - Bacterial mats operate as weirdly coordinated, arguably behaving entities using processes not too different from coordinated neurons.

Finally, certain questions can be addressed and certain myths or intellectual dodges squashed for good. Take "human awareness"—how does a dog go through a door? It does not blunder at a doorway and repeatedly attempt a series of head-butts until somehow it makes it. Why not acknowledge that the dog looked at the door, knew what it was, and went through it? Or more generally, it had some reason to go over there, remembered the layout of the building perfectly well, and knew that getting through the door was the way to go? Conversely, how does a human go through the door? Does the process involve the most edge-case, sophisticated analyses of which the human mind is capable? Do we work out geometric angles, spatiotemporal relations, the engineering of the door, and all the math of reality, using a widget perhaps, to figure it out first? No—like it or not, we go through doors pretty much the same way a dog does.

Similarly, "consciousness" as a human-referenced feature is long overdue for an unsympathetic review. Nonhuman critters can count, remember, assess, and for lack of a better word, judge the circumstances they find themselves in. Each type differs from the others in its parameters (limits, boundaries, points of focus) and in its details of individual experience, including humans. Given all that, what is left for a human-only definition of consciousness? Maybe something, but we won't figure out what it is until people stop using the term to talk about the soul with its serial numbers scraped off.

If you hold consciousness and awareness to measurable and rigorous variables, which is to say, real things, then humans have no special claim to them, and if you insist that humans have a special claim to them, then they turn out not to be real things. In my experience, people squirm when pinned down on these points, as they objectify nonhuman abilities into a grunting parody of their observed capacity, or move the goalpost of consciousness well past anything we ourselves exhibit. Because nonhumans may not think *what* we think or share it with us, it is all too easy to say they *cannot* think, or more generally, that they exhibit no cognition and that their behavior is muddy and halting, or a finely tooled automatic mechanism, anything but messy and processed thought.

It's also time to stop dodging further down the line of argument by conceding that the operations are the same in kind but insisting without justification that we do them *better*. We don't know if our physiological neural sophistication means anything about making better decisions, by any criterion, such as being right at basic problems more often, or handling more sophisticated problems in any way more successfully than basic ones. People talk themselves into this kind of claim by holding nonhumans to abstract and weird standards of "awareness" that neither we nor they can reach, and by invoking "instinct" in the sense of a colorful animal trick when a nonhuman does something smart.[4]

The word "intelligence" is so often nothing but a marker for these blind spots and refusals to engage with the question. Yes, we think and talk about doing things, to the point of obsessive chatter, and we assign social status based on personal facility with the local cultural idiom. Yet nothing ever studied physiologically or psychologically indicates that we make ordinary decisions more successfully than any creature doing any damn thing.

The same goes for emotions: nothing about the human brain and mind indicates that our intensity or complexity of feelings is anything distinctive. When pressed, the discussion dodges to the undefined term "depth" in order to maintain the exceptional human status.

Other compartmentalization tricks include focusing on conditions outside a nonhuman's processing parameters while pretending that such limits do not exist for us, and for things that are inside those parameters, cherry-picking non-human individual failures and human individual successes. However, all living functionality entails risk, as it occurs in real space and real time, riddled with conditions of failure. It's easy to forget that our "magnificent rational capacity" or nonhumans' "perfectly adapted instincts" spend most of their time making mistakes. Pertaining to this, the novel twice features a minor but significant phrase, first at the end of Chapter 13, when the Beast Folk see Prendick surrender to Moreau's and Montgomery's pleas to return to the compound, and later in Chapter 19, as Montgomery struggles to make sense of his life just before he dies.

> They may once have been animals. But I never before saw an animal trying to think.
>
> [and]
>
> "Sorry," he said presently, with an effort. He seemed trying to think. "The last," he murmured, "the last of this silly universe. What a mess—" (*The Island of Doctor Moreau*, p. 87)

Although these are Prendick's words, they express his overlap with Moreau's construct of Man/Beast thinking and intelligence. They both fail to realize that "trying to think" is the best possible description of cognitive processes as we see them in any creature.

Prendick's Gaze

Unlike Moreau and Montgomery, Prendick's understanding, perception, and judgments of the Beast Folk change during the story, several times in fact. But these changes are not from pure confusion to pure clarity, nor from pure vilification to pure acceptance—each phase is a mixed bag, as indicated by the language Prendick uses toward them.

- Chapters 3–6: He thinks they're people and is disturbed by them, but cannot tell why.
  - "[B]rown men," "black-faced man," and M'Ling and the "evil-looking" others get the personal pronoun "he."
- Chapters 7–13: He thinks they're people mutilated by surgery and is afraid to meet similar treatment.
  - The Thing, for the Leopard Man; man, Ape Man; thing, creature, Beast Folk (once), Beast Men, both "he" and "it" as personal pronouns

- Chapters 14–16: He knows they're nonhuman animals altered by surgery; he becomes increasingly more empathetic, culminating in brief but full acceptance of their humanity.
    - The Beast Folk, the brute(s), the animals, strange creatures, grotesques, the Beast People, "your brutes," "your monsters," and the appended "Man" and "Woman" to various types; "it" for many, including the Hyena-Swine; the Leopard Man is variously "he" and "it, and similarly, M'Ling is "it" but also "the black-faced man."
    - In Chapter 14, Moreau uses "her" referring to the puma, previously "it" throughout Prendick's narration, or in Moreau's comment upon her arrival, the "new stuff."
- Chapters 17–21: He knows they're nonhuman animals altered by surgery; he hates and fears them, culminating in full paranoia and murderous intent.
    - Generally, "creatures," "Beast People" (or Folk or Men), "brutes," with brute as the most common individual term
    - The transformed puma is variously "the monster," "the brute," "she," "her," and "it."
    - The Dog Man and the Hyena-Swine get the pronoun "he" in Chapter 20.
    - Chapter 21 only: Beast Monsters, with the Dog Man as "he" and the Hyena-Swine as "the monster" and "it."

The important phase is the third, when he knows the Beast Folk's origin but comes to sympathize with them more and more, even briefly to identify with them, or vice versa. This isn't what happens in most horror stories about human-like monsters, especially in groups. Let's make one up for purposes of contrast. In it, our hero, Bob, is confronted by stalking, scary humanoids—perhaps they seem like zombies, and in fact, they are zombies, actually dead, yet moving about and craving the flesh of the living. Bob doesn't realize at first that his pursuers are undead. If our story had the same number of chapters and structure as *The Island of Doctor Moreau*, then through Chapter 6, he'd think they were crazy humans, and then in Chapters 7 through 13, he'd think they were living people who were altered in some way. Throughout these chapters he might find them scary and upsetting, but still maintain some sense of sympathy for their plight. But then, in the transition between Chapters 13 and 14, he'd correctly learn that they're classic Hollywood undead.

Well, that does it. Bob goes into full ruthless mode for the rest of the story, shooting the zombies in the head (and similar actions) in order to keep from being eaten, and I think you know how the story goes from there.

In this story, the plot includes a single, perfectly symmetrical flip between Bob's knowledge and his empathy: as soon as he shifts from fully mistaken to fully informed, his initial sympathy disappears. He might conceivably retain a certain grief or smidgeon of personal horror after that, but not enough to affect his trigger finger; in fact, those secondary characters who do let their empathy interfere with their ruthlessness would be depicted as fatally naïve.

However, in the novel, Bob's story isn't what Prendick goes through at all. The transition from mistaken interpretation to complete knowledge is the same, shifting between Chapters 13 and 14. Also, his initial empathy levels are similar in the first two phases as well: disturbed and then horrified regarding what he thinks are distorted human beings. However, when his knowledge flips to the accurate understanding that their origins are fully nonhuman, his empathy levels go through two extreme transitions that map very differently to the knowledge transition.

Since Prendick's emotional through line in the story does not hinge strictly upon his knowledge of the Beast Folks' nonhuman origins, it cannot be simplistically slotted into good versus bad categories defined by a man versus beast divide. During Chapters 14 through 16, after he knows what the Beast Folk are, he steadily feels more and more positively toward them. His response begins mainly as sympathy for tortured animals, but then acquires genuine empathy, which peaks during the events of the hunt for the Leopard Man at the end of Chapter 16. At that point, he acts upon and articulates the most uncompromising and least sentimental embrace of nonhumans as fellow beings in literature.

But right at the beginning of Chapter 17, after the mostly undescribed six-week gap in his account, this feeling is entirely gone. Prendick has apparently flatly reversed his perceptions and insights of Chapter 16. He even displays a new distaste and loathing for the Beast Folk, more extreme than his initial disturbance upon originally meeting them. Right at the opening of the chapter, he describes the Beast Folk as "horrible caricatures" of real people, and describes real people as assuming "idyllic beauty and virtue in my memory."

The textual transition is so abrupt and extreme that it comes off as uncharitable of him, and it instantly begs the question of how his shift in judgment developed during the undescribed six weeks of his stay upon the island. However it happened, it's permanent; he never again expresses sentiments like those which closed Chapter 16. Instead, his distaste continues and becomes more intense, in a complex process intertwining religion and the Beast Folks' reversion.

This profile of understanding and judgments captures the two core questions of the plot:

1. Why does Prendick come to feel so deeply in favor of the Beast Folk's humanity *after* he learns what they are?
2. Given his deep empathy for them at the end of Chapter 16, why does he come to hate and fear them so dramatically by the start of Chapter 17?

"No!"

The moment when Prendick's assessment of the Beast Folk hits its most positive, most sympathetic level is wrapped around the subplot of the Leopard Man, culminating in Prendick killing him. It begins in Chapter 9 with the two as antagonists, just before the Leopard Man stalks Prendick through the forest, probably barely suppressing the urge to take him down as prey.

It's a dark, mysterious scene, reflecting Prendick's complete confusion. I'm not sure what the Leopard Man means by saying, "No!" when they come face to face—is it a preemptive refusal of whatever Prendick may order? Is it in defiance of his conditioning, therefore an expression of his desire to kill? Or is it an instruction to himself, to reinforce his understanding that he should not kill prey? It's like a human's behavior, but the reader can't be sure if it is.

I think the mystery arises because the Beast Folk experience complex emotions although their creators do not see it, or they don't admit what they do see. For example, Montgomery unwittingly reveals quite a bit when he states that he and Moreau must *prove* (emphasis in text) to the Beast Folk that the Leopard Man killed the rabbit, in order to punish him. Moreau confirms this necessity soon afterward. In other words, the Beast Folk possess a perfectly sound concept of evidence-based jurisprudence and are capable of critiquing the claims of authority, not only individually, but collectively. Moreau and Montgomery cannot simply invoke the Law and crack the whip. Although they both know this and casually refer to it, they don't *notice* that they know.

Prendick is not sympathetic toward the Leopard Man, to say the least, and in Chapter 16 he joins in his exposure and in the ensuing chase with a will. Wells clearly did his share of running around in rough terrain; I have, too, and I can confirm that the extended chase sequence perfectly captures the changing ground, the painful and inconvenient struggles through the underbrush, and the light-headed combination of enthusiasm and exhaustion. In such moments, emotions appear suddenly in a sharp, distinct way.

In the story, two such emotions stand out. One is Prendick's distaste for the presence of the Hyena-Swine, who is clearly enjoying his opportunity to pin the crime and upcoming punishment solely on his accomplice, as well as sizing up Prendick for some further act. He reeks of hypocrisy, glee, and malevolence, perfectly displaying the stereotypes of mockery and profanity invoked by the animals he's made from. However, those stereotypes have nothing to do with real hyenas or pigs but with people. Prendick despises the Hyena-Swine because he quickly and accurately *recognizes* his behavior.

The other is just before they find their quarry, when Prendick's feelings for him change dramatically:

> "Back to the House of Pain, House of Pain, House of Pain!" yelped the voice of the Ape-Man, some twenty yards to the right.
>
> When I heard that, I forgave the wretch [*the Leopard Man*] all the fear he had inspired in me. (*The Island of Doctor Moreau*, p. 72)

When Prendick locates the Leopard Man before the others do, this feeling culminates in this crucial phrase:

> It may seem a strange contradiction in me—I cannot explain the fact—but now, seeing the creature there in a perfectly animal attitude, with the light gleaming in its eyes and its imperfectly human face distorted with terror, I realized again the fact of its humanity. (Ibid., p. 72)

Victorian-era writers did not screw around with their word choice. It doesn't say *semblance of humanity, akin to humanity,* or *my impression of its humanity,* or anything like those. It doesn't even say humanity by itself, open to interpretation. Prendick abandons all such qualification—the word is "fact."

What is human, at this point in the story? It's explicitly not a being's bipedal posture, or the possibility of its being mistaken for a member of our species, physically speaking. It is instead the Leopard Man's understanding of his situation and his multiple resulting emotions: his recognition of being discovered, his terror of further torture, and what can only be called moral despair. Significantly, these fall into the category of the things Moreau earlier claimed he could *not* produce through his techniques, but which are now revealed to Prendick to exist, in this person, in a form he identifies with easily.

Whereupon Prendick, knowing that Moreau wants to return the Leopard Man to the laboratory both as an experiment in reprogramming and to demonstrate to the other Beast Folk that "none escape," shoots him between the eyes. He does so from *sympathy* to prevent the suffering of a (i.e., any) living thing, but also from *empathy,* to relieve a fellow man of despair, both at that moment and in his inevitable agony later in the laboratory.

What mental operations make this possible? It lies in communication, which strictly speaking, is a carefully defined, social subset of perception. It happens when one individual perceives the signals of another, and acknowledges to the sender that they have received them. This behavior is very, very common among social organisms, employing a huge range of media and senses, of which the cognitive trick of language and the physical medium of vocalizing are subsets.

The real barrier to insight is our experience of language, which is all too easy to inflate in value, creating a circular argument. "We speak because we have special mental abilities, and we think in a special way because we speak." Technically, language is indeed a neat trick: it folds two things together, vocabulary and syntax, with surprising results that neither can do alone. One such result in our species is recording content, so that older communications may become a teaching curriculum in later generations, which we call "culture." It's definitely distinctive as a historical combination of things, but not a new thing at an atomic level, because both vocabulary and syntax are observed across other living species.[5]

However, we make too much of it in several ways. For one, we consider the vocal medium that we use for this communication as the phenomenon itself, flying in the face of profound evidence that the use of sound has nothing intrinsically important about it toward such a function. For another, the recorded form of language has had drastic ecological consequences, not least our recent population explosion, but in this, it's similar to many accumulated and multifaceted biological phenomena in many species. This discussion always boils down to identifying our population explosion as an achievement, a heroic triumph, a debatable point to say the least.

So I'll break the circle. Like many other primates, humans are extremely good at the social form of cognition, permitting them to evaluate who is doing what to whom at many levels and across many individuals. Our map of social community is, perhaps, the central organizing principle of human cognition, such that we can hardly imagine "communication" without it. If so, then misidentifying our mode of communication, speech, as the cognition itself, or pulling the trick that all other forms are substandard because they do not use it, is a mire nearly impossible to escape. It means that we confound the capacity to communicate with membership of an acknowledged community and even with human status. Sadly, we constantly demonstrate this in the demonizing of other people simply due to differences in human language.

Understanding human cognition, then, lies in our particular social operation rather than the linguistic one. Work on social cognition has vastly improved in the past decade, but even some of the best discussion cleanly misses the crucial question: Does human behavior meet the criteria for the various goalpost-shifting concepts of consciousness? The discourse always takes it as given that it does, and as I see it, dives back into the intellectual hamster wheel encountered by so many of us who are interested in this topic.

That's not to discount the research insights. One useful concept that's emerged is called the language of thought, to investigate social and cognitive operations without getting distracted by verbal speech. The first insight is to recognize how widespread it must be, and how it might be worthwhile to consider other complex variables, like communicative devices besides vocalizations in other social animals, and equivalently complicated environments for non-social ones.

Social cognition and the suddenly shared language of thought are at the core of this moment in the story. However briefly, the whole situation suddenly becomes not Prendick's place to say something, but the Leopard Man's. Until this point, the Beast Folks' ability to vocalize words had not by itself penetrated Prendick's sense of shared community; it was not communication. Now, however, he cannot interpret this immediate situation in any other way as he acknowledges common values, fears, perceptions, and concepts. He's not projecting. It's not a guess. He "gets" the Leopard Man *socially*, and once the connection is made in that medium, then by definition, the two can communicate. The Leopard Man's unvoiced statement strikes Prendick between the eyes as surely as the bullet will strike him in return: *Please don't let them take me.*

Such an insight is life transforming. Prendick quickly translates his new grasp of the Beast Folk's nuanced and above all familiar experience of their lives into equivalency between them and humans as he knows them. As he soon puts it,

> A strange persuasion came upon me that save for the grossness of the line, the grotesqueness of the forms, I had before me the whole balance of human life in miniature, the whole interplay of instinct, reason, and fate in its simplest form. (*The Island of Doctor Moreau*, pp. 73–74)

This is as far from the "don't meddle" science fiction horror story as you can get. Prendick is briefly unconcerned with the creations' potential for violence ("Oh no, they're really animals, they'll turn savage and eat us!"). Instead, his sympathy to their physical plight as nonhuman victims of torture expands dramatically to empathy with their emotional, intellectual, and spiritual plight as people like himself. He isn't even patronizing about it—it's nothing to do with the Leopard Man being *as good as* a person, but rather that the *totality* of personhood and the Leopard Man, with all his faults and fears, are the same.

A lot of "don't meddle" stories cast the subject of the abominable experiment as a tragic victim, especially when it's rejected: the creature is angry because it is different, because it is cast out, because it is isolated and unloved. That's not what happens in this novel. The Beast Folk do not rage and revolt—they putter along, living their lives, getting married, and growing their vegetables. They stumble frequently in the eyes of the Law, and try to make the best of it. Their failures and grotesqueries are *ordinary*, not fantastic results of their fantastic origins.

> The Leopard Man happened to go under. That was all the difference. (Ibid., p. 74)

It wasn't his Animal urges, beneath our lordly status or our Man or Angel side, that dragged him under "back to the beasts," but plain old urges, any urges, such as you or I might feel and similarly struggle with, and which dragged him exactly to the emotional and social crisis where you or I might well go. He did a bad thing a person might do. Looking back over his story, I think the Leopard Man's "No!" is not evidence of his alleged bestiality, but rather the defiant cry of someone who can no longer stand feeling guilty for something he did, yet still feeling it, who will rebel even harder if he's criticized.

Prendick may be forgiven his insensitivity to these nuances, perhaps, as he began their interaction with being stalked and then attacked in the forest at night by a man who is also a leopard. It could be said that this initial moment was pure rotten luck for both of them, as given a chance, Prendick might have listened to him. As it went, though, he failed to see the community-and-communication medium they shared, and realized too late that this other person might have something to say.

Two of the films make use of this subplot, but they both distort its core features to the point of incoherence.

In *The Island of Doctor Moreau* (1977), the sequence is split into three characters, including M'Ling as the sucker-up of drink, who is then reprogrammed in much agony, then a cat-like, possibly tiger-based man who chases Braddock through the forest and attacks him on sight in defiance of the Sayer of the Law, and who is similarly reprogrammed, and finally the bull-man who defies the Law in the name of "Animal! Proud!" and goes on a rampage. The two violent rebellions seem to be expressed as a random, incomprehensible desire to fight the most dangerous opponent available for no reason, in the latter case an unmodified tiger that has escaped its cage. Spectacular and terrifying as it is to see a stuntman wrestle an actual tiger (no CGI back then!), it makes almost no sense at all. After another group confrontation with Moreau, the bull-man flees in fear of the House of Pain, chased by everyone. His final scene retains some of the pathos in the novel, when he pleads for Braddock to shoot him, "No House of Pain . . . kill," which he does. The lack of prior interaction between them, though, leaves it as a toss-up between a moment of genuine connection between two people and the more familiar, "don't meddle" concept of "We belong dead" as realized by a wretched, science-created monstrosity. The idea that the bull-man is more human than Braddock thought is at least possibly present in his understandable desire not to be painfully brainwashed, but is also undercut by his expression of "Animal!" as an urgent and apparently constant desire to fight.

The corresponding events in *The Island of Doctor Moreau* (1996) at first accord slightly more with the novel, featuring a leopard-man named Lo-Mai (Mark Dacascos) who kills a rabbit, the confrontation between Moreau and the Law-abiding Beast Folk community, the hyena character who hides his complicity, and the ensuing brief attack on Moreau. However, Douglas has no relationship with Lo-Mai and plays no role whatsoever in his eventual fate, when even as he submits to Moreau, he is unexpectedly shot by Azazello, the dog-man. The latter is obviously smugly pleased with murder, has inexplicably been given a gun by Montgomery, and is equally inexplicably not punished or limited in his later activities in any way. This point in the film marks its veering into many unmotivated or logistically hard-to-follow events, and whatever intellectual content or emotional resonance might have been evoked is lost.

The Virago

In terms of story pacing and plot significance, the insight into the Leopard Man's situation is as fleeting as Prendick's perception of it. There's another, prolonged, step-by-step point of view, with considerably more impact on everyone in the story, if you're not blinding yourself to it. It belongs to the puma (in the US, mountain lion) who arrives on the island with Prendick.

She is mentioned and described throughout Chapters 3 through 8, from Prendick's recovery aboard the *Ipecacuanha* through his acceptance by Moreau, so it's possible to trace her experiences in detail—most of them horrible, such as being cramped into her transport cage, or spun around by the tackle while being unloaded. Once arrived, her surgeries begin immediately. She doesn't appear in Chapter 9 only because Prendick has left the compound to escape her screams, which are described so eloquently that when they reappear in Chapter 10, I find that I have retroactively extended them back through Chapter 9 even though Prendick couldn't hear them.

Her voice changes in Chapter 10, after only a day or two of surgery, to the extent that Prendick is convinced that Moreau is operating upon a human person. In Chapter 14, Moreau speaks of intervening extensively into her brain, occasionally so distracted by his hopes for success (again, for a "rational being") that he sometimes loses his train of thought when debating with Prendick. Whatever degree of human mental processing is developed by Moreau's techniques, which as I argue in this chapter is considerable, is well along its way. It's also at this point that the reader learns her gender.

By the opening of Chapter 17, after the six-week break in Prendick's narration, she has been subjected to almost two months of constant and extraordinary agony, permitted to heal between sessions only to be set upon again. When Moreau enters his laboratory, she shrieks, and here the text employs an interesting term:

> So indurated was I to the abomination of [*Moreau's laboratory*], that
> I heard without a touch of emotion the puma victim begin another day of
> torture. It met its persecutor with a shriek almost exactly like that of an
> angry virago. (*The Island of Doctor Moreau*, p. 75)

Some of the criticism I've read focuses on the negative implications of the word, similar to "bitch," "hysterical," and "uppity," but I also consider its older, original meaning: a woman of strength, assertion, presence, and force, willing to defy social norms without denying her sex.

What happens then is simple: she does what no other Beast Person has done, even those made from animals probably stronger than she is, such as the oxen or the bear. Either she pulls her chain out of the wall upon seeing Moreau, or she did so during the night and has been waiting for him. She hits Moreau with it—not employing her teeth or claws—and breaks Prendick's arm on her way out of the compound, then runs onto the beach. When she perceives Moreau pursuing her she takes to the brush, escaping him briefly.

The confrontation between her and Moreau must be reconstructed from what Prendick and Montgomery find almost a day later. When Moreau caught up to her, he shot her through the shoulder, but she was still able to kill him, again using the chain. She collapsed from loss of blood from the bullet wound, and she either bled to death or was too weakened to defend herself from the other Beast Folk. Her body is found "gnawed and mutilated."

Several authors have investigated the character regarding fear of women and their oppression by Victorian and later society, especially in light of the prevailing ideas that women are more emotional and therefore more animalistic than men, unless subjected to male authority, up to and including surgical intervention. I agree with and recommend Coral Lansbury's discussion of that issue in *The Old Brown Dog*, but I'm focusing instead on the character's active and decisive role, rather than solely her victimization.

The Puma Woman, as I shall call her, is an actor, not merely acted upon. She extends the Leopard Man's relatively passive defiance, becoming more than a skulking victim and sinner. Consider her experience just prior to the escape. What did she know? What did she think? How long did it take to get that chain out of the wall—was it a sudden surge of adrenaline, or had she tried all night—or even, possibly, over many nights?

Her cry, "almost that of an angry virago"—isn't it the most easily identifiable moral voice in the story? Her agonized screams provided the backdrop to Moreau's and Prendick's debate in Chapter 14—now, isn't this new cry, delivered not in agony but in anticipation, effectively her decision to participate in that debate herself?

Here the story taps into the issues of justice that Moreau's and Prendick's debate missed entirely. She has something to say, or to do, precisely about what is being done to her. Or to put it rather more bluntly, what might the vole have said to me?

I've criticized the movies a lot, but I'll give their creators credit for examining the puma's potential as a major character, especially since they don't merely glamorize and objectify her. Her decisions and sexuality are a serious plot point begging to happen, and the films do come through in varying ways, generating deep ambiguity regarding whether this character "should" exist and given that she does, what degree of empathy, attraction, and admiration she merits, and most especially, what she chooses to do.

In *Island of Lost Souls*, in addition to the unscientific craziness I discussed previously, Moreau is even more perverse in his sexuality, always expressed through some hypothesis test or other. He sets up Parker and Lota for romance—successfully!—and lurks in the shadows to watch them kiss, then he sets up Parker's fiancée Ruth to be raped by Ouran, all in the name of testing how human his creations have become. Film writers have a field day with him, including the explicit sublimation (no one ever looked more like an impotent voyeur than Laughton's Moreau) and the racist stereotyping of the exotic, innocent native woman who wants nothing more than to keep the mighty white man by her side, and the bestial, grinning native man, who needs only glance at the white woman to become obsessed with raping her. The literature on that matter is fascinating, but my aim is a little different.

Although this film's plot is the pure, even distilled "don't meddle" story, Lota's role in it is deeper, including some positive elements of the novel's Leopard Man and Hyena-Swine, aided by a remarkable performance by Kathleen Burke. Her moral status, which is to say, how to think of her, is dramatized by the escalating romantic triangle among Lota, Parker, and Ruth. The triangle is resolved orthogonally, rather than through a direct conflict, when Lota sacrifices herself to save the others from Ouran. Ruth never learns about the knee-weakening kiss, for instance. I was a little disappointed not to see Parker forced to choose between the two women, which would entail deciding whether Lota is human or not. Her final actions spare him that choice even as they nail that very issue to the wall—in her favor.

In *Terror Is a Man*, the romantic tangle is effectively the whole story, with Girard's wife Frances at the center of all four principal men's interests, including her husband as a driven but sane version of Moreau and the nearly complete Panther Man. The gender switch doesn't change the issue, because the Beast Man turns out to be rather admirable toward her, especially in contrast to Walter the debased assistant, and even in contrast to Fitzgerald, the protagonist. As in *Island of Lost Souls*, the sexualized and emotionally sincere Beast Person is sympathetic, but in this case, as the only Beast Person in the story, he's also the source of lethal danger to others every time he gets loose. Fitzgerald's opinion of him captures that dilemma, as he states that upon looking into the Panther Man's eyes, he sees a soul, but he also fears the created man's evident urge to kill—as if a tormented, restrained person would not return a captor's scrutiny with the desire to strike back.

The Panther Man's final confrontation with Girard is almost exactly the same as the Puma Woman's with Moreau in the novel, and his death seems quite sympathetic to me. I can't help wishing that his murderous attacks on the natives had been left out of the plot, such that his lethality would have been limited to his tormentor Walter. In that case, the story's rather well-developed question, "what makes a man," might have been thrown into unique focus through Frances's eyes.

The main conflicts in *Twilight People* are also rooted in romantic tension, although in this case, only among three fully human characters and with one corner being explicitly homoerotic. The Panther Woman, Ayessa (Pam

Grier), isn't part of this story at all, so adds little to my present point, aside from being the film's unquestioned most righteous bad-ass and killing a slew of Doctor Gordon's goons. Her presence accords more with Jennifer Vere Brody's analysis in *Impossible Purities*, as the films, especially this one, generally go further than the novel ever does in exoticizing the Beast People in ethnic terms. My point is still supported, though, in that Gordon meets his end at the hands of his wife, now severely transformed by his experiments.

In *The Island of Doctor Moreau* (1977), Maria (Barbara Carrera) is presented such that she may or may not be one of Moreau's creations, despite many hints, and she and Braddock do become lovers. However, this plot thread or question becomes a bit lost in all the mayhem until the very end, where I find myself muddled slightly—because I would swear on a stack of *Metropolis* DVDs that when I first saw this movie in my mid-teens, Maria is clearly reverting to her panther form in the final moments, in perfect mirror to Braddock's recovery from Moreau's animalizing serum.[6] I was looking forward to seeing the final scene again, but on the currently available DVD, what do I find? The final shot of Maria's face isn't at all what I remember, but that of an ordinary woman with maybe the barest touches of makeup to suggest otherwise, if that. Was she a creation of Moreau's or wasn't she? Answering "no" makes no sense, considering multiple lines of dialogue and crucially, that Moreau was shown to be observing the earlier sex scene in a clinical fashion, but the visuals lean more toward "no" than "yes."

As the single unfortunate exception, in *The Island of Doctor Moreau* (1996), Aissa (Fairuza Balk) is deeply marginalized by the script, which makes her less of a positive virago and more of a waif. She's a cat, not a puma, and her action is limited to wavering between ineffectually helping Douglas and irrelevantly tending to Moreau. Even her catlike qualities are weak when she needs them, as she gets beaten up easily by the dog-man, and she ultimately ends up the victim of trying to please too many people at once—not exactly the route the original puma character or the other film versions of her would take.

At the other end of the spectrum is *Doctor Moreau's House of Pain* (2004), in which Alliana (Lorielle New, mis-credited as Loriele New) is—"objectified" is about the mildest way to put it, as she works as a stripper, which is filmed as to leave no doubt, seeks a suitable mate, ditto, and deals with an annoying man by punching her fist through his head. However, there's no question that the film is also working with motifs relevant to what I'm discussing in this section. She and the other transformed animals in this film hold a lot of agency, making decisions as least as important as those of the human characters, and New's performance is similar to Burke's in *Island of Lost Souls* in combining catlike movement with undeniable personhood. Alliana's final decisions are more vicious and self-centered than Lota's, but they cannot be described as brutish or subordinate; for instance, unlike Lota, her sexual and romantic focus on Carson is not ordered or manipulated by Moreau but completely under her control.

Refusing to Look

Prendick's changing gaze culminates in retreat, ending in utter exceptionalism and, in the face of experiences that contradict it, break down. It begins with that full and sudden flip that starts in Chapter 17. After six weeks pass, about which Prendick reveals almost nothing, he has adopted Moreau's idealized view of Man and actively suppresses the fact that the latter's words criticizing the Beast Folk inadvertently describe people perfectly, or that he had seen the fact of their humanity for himself.

> . . . I had lost every feeling but dislike and abhorrence of these infamous experiments of Moreau's. My one idea was to get away from these horrible caricatures of my Maker's image, back to the sweet and wholesome intercourse of men. My fellow creatures, from whom I was thus separated, began to assume idyllic virtue and beauty in my memory. (*The Island of Doctor Moreau*, p. 75)

Weird! I would think, after his insights at the end of Chapter 16, that he'd go straight back to the lab and insist that the Puma Woman be freed. But he doesn't, and after six more weeks of listening to her scream, he's "untouched by emotion" and coming up with all this hate speech.

He's also avoiding Montgomery like the plague, and no wonder: the assistant is all too evidently a person, rather than a paragon of virtue and beauty.

Prendick never found his way into a gaze that wasn't his own, which is why he failed to help the Leopard Man in time. As for the Puma Woman, he fails to help her at all, completely missing her presence in the situation, despite it being his primary sensory experience throughout most of the first half of the book. The reason why is twisted up in that strange six weeks of silence between two chapters, and what can only be interpreted not merely as happening to miss her, but refusing to look.

The power in this portion of story arises, at least as I see it, from forcing the reader—you and me—to look instead, to challenge the Man/Beast divide in precisely those places we cling to it: our "intelligence," our "awareness," and our over-vaunted verbal tricks. It's laid upon us to make the connection that Prendick could not.

The literary power of that point is stunning, even historically heroic given the period, and given that the same conundrum ties our intellectual culture in knots. It gets further than Huxley did, and further than Wells would achieve again.

Readings

James Rachels, *Created from Animals* (1999), provides a good summary of Darwin's work on nonhuman decision-making and problem-solving. The definitive reference is Robert J. Richards, "Darwin and the Emergence of Evolutionary Theories of Mind and Behavior" (1987).

Academic writing on human thought is contentious, so the following titles are recommended as an assemblage of slightly differing views and as portals to their extensive bibliographies: Marc Hauser, *Wild Minds* (2001); Mark S. Blumberg, *Basic Instinct* (2005); Edmund O. Wilson, *On Human Nature* (1994); Stephen Pinker, *The Language Instinct* (1995) and *The Blank Slate* (2003); Timothy Goldsmith and William Zimmerman, *Biology, Evolution, and Human Nature* (2000); James Gould and Carol Gould, *The Animal Mind* (1994); Dorothy Cheney and Robert Seyfarth, *Baboon Metaphysics* (2007); and Temple Grandin, *Animals in Translation* (2004). Technical mind-and-brain function references include M. Deric Bownds, *The Biology of Mind* (1999); and John Dowling, *Creating Mind* (1999).

Intersections among clinical psychotherapy, social deconstructionism, and literature can be found in Paul Gilbert (editor), *Compassion: Conceptualisations, Research, and Use in Psychotherapy* (2005), especially Chapter 8: Neville Hood, "Cosmetic Surgeons of the Social." Further historical and gender analysis are available in Jennifer Vere Brody, *Impossible Purities* (2012), and Coral Lansbury, *The Old Brown Dog* (1985).

The Elsevier *Trends* journals published by Cell Press are one of modern biology's brightest lights, publishing reviews and opinion pieces in the hope of bringing current debates to the attention of multiple disciplines, highlighting unusual points of view, and sparking new questions, and best of all, many articles prompt fierce debates in subsequent issues. There are fourteen titles, including *Trends in Ecology and Evolution, Trends in Cognitive Sciences*, and *Trends in Molecular Medicine*. Most universities have licensed them for free reading online, so if you have access, I can recommend no better source to keep up with what bioscience and biotech are doing at the edge of what we don't know.

Notes

1. The connection with Freud's writings is direct: the id is defined as repressed stages of thought from older evolutionary conditions, primitive predispositions from prior organismal identity.

2. Credit for this one goes to Leon Eisenberg, "Mindlessness and Brainlessness in Psychiatry," *British Journal of Psychiatry* 148(5): 497–508, 1986, although, as his title states, he was discussing clinical care, not the broader behavioral and evolutionary field.

3. *Trends* references:
   - Regions of the cerebral cortex
     - Steve L. Franconeri et al., "Flexible Cognitive Resources: Competitive Content Maps for Attention and Memory," *Trends in Cognitive Sciences* 17(3): 134–141, 2013.
     - Kalina Christoff et al., "Specifying the Self for Cognitive Neuroscience," *Trends in Cognitive Sciences* 15(3): 104–112, 2011.
   - Cell functions within the cortex
     - Matthew Larkum, "A Cellular Mechanism for Cortical Associations: An Organizing Principle for the Cerebral Cortex," *Trends in Neurosciences* 36(3): 141–151, 2013.
     - Anis Contractor et al., "Kainate Receptors Coming of Age: Milestones of Two Decades of Research," *Trends in Neurosciences* 34(3): 154–163, 2011.
   - Remodeling and plasticity
     - Hiroaki Wake et al., "Microglia: Actively Surveying and Shaping Neuronal Circuit Structure and Function," *Trends in Neurosciences* 36(4): 209–217, 2013.
     - Min Fu and Yi Zuo, "Experience-Dependent Structural Plasticity in the Cortex," *Trends in Neurosciences* 34(1): 177–187, 2011.
   - Hippocampus and memory
     - Gianfranco Dalla Barba, "The Hippocampus, a Time Machine That Makes Errors," *Trends in Cognitive Sciences* 17(3): 102–104, 2013.
     - Sara A. Burke and Carol A. Barnes, "Senescent Synapses and Hippocampal Circuit Dynamics," *Trends in Neurosciences* 33(3): 153–161, 2010.
     - Howard Eichenbaum, "Memory on Time," *Trends in Cognitive Sciences* 17(2): 81–88, 2013.
     - Oliver Hardt et al., "Decay Happens: The Role of Active Forgetting in Memory," *Trends in Cognitive Sciences* 17(3): 111–120, 2013.
   - Bacterial mat signaling
     - Robert P. Ryan and J. Maxwell Dow, "Communication with a Growing Family: Diffusible Signal Factor (DSF) Signaling in Bacteria," *Trends in Microbiology* 19(3): 145–152, 2011.

4. The word "intelligence" lacks scientific content and has probably been corrupted permanently in its use as a performance variable in stress-testing that we call IQ tests. If it were to be rehabilitated, it would certainly not include that type of performance. See Stephen Jay Gould, *The Mismeasure of Man* (1982).

5. Vocabulary is a catalog of representations used in communication, "words" in the general sense; one's vocabulary is typically acquired through examples. Syntax is a cognitive framework of rules or techniques for constructing sentences; it underlies grammar and makes language as we know it possible. The honeybee provides an excellent example of a nonhuman that uses syntax. It may be more useful, however, to consider communication as such before language as structure, as demonstrated by Uri Hasson et al., "Brain-to-Brain Coupling: A Mechanism for Creating and Sharing a Social World," *Trends in Cognitive Sciences* 16(2): 114–121, 2012.

6. I remember speculating that it would have been awesome if Braddock had died in the final fight, and if the approaching ship's crew would have therefore discovered nothing in the boat but a panther.

7

To the Beasts You May Go

Deep in the story lie those six undescribed weeks, soon followed by Moreau's and Montgomery's deaths. At the start of Chapter 17, already refusing to look at the Beast Folk, Prendick is now refusing to look at his fellow humans, both immediately and in his memory of his ordinary life. In Chapter 22, back in England, Prendick's ending point of view has retreated further and completely into exceptionalism. As with Montgomery back on the island, he quickly removes himself from contact with ordinary people. People are the last thing Prendick wants to look at or deal with, because he sees that humans are already, and always have been, the Beast Folk.

> I see faces keen and bright, others dull and dangerous, others unsteady, insincere; none that have the calm authority of a reasonable soul. [. . .] I know this is an illusion, that these seeming men and women about me are indeed men and women, men and women forever, perfectly reasonable creatures, full of human desires and tender solicitude, emancipated from instinct, and the slaves of no fantastic Law—being altogether different from the Beast Folk. (*The Island of Doctor Moreau*, p. 103)

Are real people emancipated from instinct? Subject to no fantastic law? Of course not, as Prendick saw so clearly at the end of Chapter 16. However, this late-stage Prendick has now completely adopted Moreau's view, and in reporting what he cannot help but see about his fellow humans, he might as well be quoting Moreau's lines about the Beast Folk in Chapter 14:

> I can see through it all, see into their very souls, and see there nothing but the souls of beasts, beasts that perish—anger and the lusts to live and gratify themselves. [. . .] There is a kind of upward striving in them, part vanity, part waste sexual emotion, part waste curiosity. It only mocks me. . . . (Ibid., p. 59)

Just as Moreau hints here, Prendick is not happy in holding this view—too much has occurred for his compartmentalization to work, and he's writhing in the full-on stress of cognitive dissonance. His insistence upon exceptionalism is a (failing) defense to keep from facing his own conclusions, but he can only idealize humanity this way as long as he doesn't actually have to deal with a living human being.

What does it mean to look at ourselves as humans? In the late nineteenth century, the scientific and social texts were largely engaged with human achievement or special status, a conviction that informed the origins of both anthropology and human paleontology and which neither discipline has quite managed to shake off. Later biologists, even those who contributed immensely to our understanding of mammalogy and behavior today, kept a hands-off policy toward humans, passively reinforcing the same idea. Still, every so often and in many ways distorted by individual quirks or instant interpretations, different perspectives have arisen, at least among some of us, beginning with the notion that "people are animals," and running into crisis because, even then, it's so hard to ask, "So what?" without expecting a really profound answer.

Maybe that question will be important some day. But we jump to it too fast, with too few solid observations or well-analyzed conclusions, with too much drama still riding on it, and without useful vocabulary. A good scientific approach to this issue wouldn't be expected to provide wonderful new answers to something-or-other. It should instead, like every good new model, provide better, more convincing explanations for a wide variety of things that we already know.

No Threshold

Getting your head or anyone else's into this space runs up against two layers of defense, first the "magic free thinker" one, about what's inside our heads, which I hope to have dealt with in Chapter 6, and then the harder one: no longer explicit spiritual exceptionalism, which can at least be identified, but exceptionalized biology, which is more subtle and insidious. It comes in two layers: the claims that sound like or purport to be biology but aren't, and worst of all, twisted versions of biological ideas that persist in the discipline.

The argument runs, perhaps there's no solid cutoff between humans and other animals in physiology or cognition, but are there not effects that have emerged which unequivocally show that we are special in some way? What about the physical evidence of progressive change, our sole representation of our type of creature, the achievement of our species—is there not a gap there to be seen easily?

The answer is no. We aren't who we think we are.

How We Got Here

The first insight from the larger evolutionary history is that so much of what we identity as "us" did not evolve within or just prior to *Homo sapiens* at all. Instead of "humans, unlike other animals," consider "humans, characteristic of this type of animal."

Here we are (Figure 7.1), genus *Homo* with its known species and its known close relatives. Consider the extinct forms and phylogeny of this subset of apes. The first thing you notice, I hope, is that they're all standing up. The hunched-over caveman never existed; that image is an artifact of three things: the fraudulent fossil (which wasn't even a fossil) called Piltdown Man, the persistent error

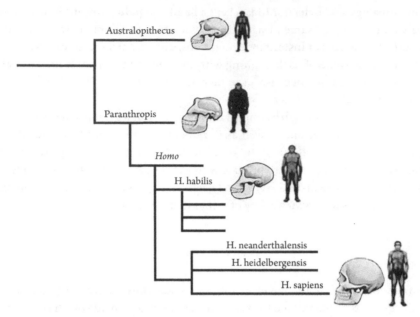

Figure 7.1 Phylogeny with silhouettes.

in twentieth-century anthropology to think that every piece of the body evolves simultaneously and gradually, and a lot of pop culture.[1]

Let's convert that branching diagram into the more accurate if more cumbersome nested-boxes version. Here the answer is clearer: we got our taillessness and tooth row count from being apes, we got our bipedality, thumbs, and reduced canine teeth from *Australopithecus* or something very much like it, our big-brain template from *Homo habilis*, our capacity to disperse from *H. erectus*, and most of our extremely big brain from the common ancestor of *H. sapiens, H. heidelbergensis,* and *H. neanderthalensis.*

Bluntly, there is no identifiable point for a definitive stamp of humanity—not from any single feature (and *my God* have people tried!), not at any single point in the process, nor as a uniform subtle change simultaneously across all the features. Instead, you can see multiple different origins for many features that we currently see all at once and mistake for an artificially unified thing. This is what always happens when you try to understand the origins of a relict species, which is to say, one or a very few species that still exist from a previously much more diverse, but now mostly extinct assemblage of species.

Consider the eastern Australian platypus (*Ornithorhynchus anatinus*), for instance: this fuzzy mammal has a bill similar to a duck's, features spurs on its ankles that are venomous in males, lays eggs, and secretes milk through an area on its skin, without teats. As a whole unit, it looks completely made up, and not unreasonably was suspected to be a hoax when first reported in Europe in 1798. In biological terms, this or any number of weird animals leads one wrongly to infer that the evolution that produced it must have been correspondingly unusual. A person might start thinking about strange environments, assuming that all its distinctive parts evolved together and were confined to this one species or small group of species that we see. However, these responses are an artifact of our limited observational processing—we think that what we see is a "thing" equivalent to other "things" we know—but it's not. The platypus and the few similar mammals, collectively called monotremes, are the last few remaining species of a previously very large and diverse assemblage of animals. Their various funny-looking bits didn't evolve all at once, but at different times throughout this larger, now invisible phylogeny, and no special evolutionary history need be inferred—merely an evolutionary history, period. In fact, the living forms are not particularly closely related and may not even comprise a single nameable group.

Homo sapiens is also a classic relict species, although much less dramatically so than the platypus. We're the only extant species from a group of over twenty kinds of bipedal apes in several distinct groups. And sure enough, we're subject to precisely the same critical failure in assessing our evolutionary history. We're a biped! With a big brain! Teeth like this, hair like this, and so on and on—it must all be connected, right? Evolved in a special way, right?

To correct this failure, the operative concept is called mosaic evolution, so named because a mosaic image is composed of many small tiles, each of which can be independently altered or traded out. Once you include the whole phylogeny with its extinct forms, you can see that almost none of our profile of interesting parts evolved together.

Unfortunately, the bipedalism and freakishly inflated head are all we've got from the paleontology. Limited as that is, it's enough to demolish the long-standing portrait of evolution as a series of terminal additions, in which each step "gets" a thing, and the later steps therefore have more things with necessarily more complex interactions among them. That picture of the creatures walking along from left to right, showing a more-and-more human progression, used in a million memes? Forget it.

Perhaps you noticed someone unusual in the phylogenetic diagram, always conspicuously absent in that misleading image. It's *Paranthropus*, the perfect case study for mosaic evolution because it shares our early ancestry but changes very different "tiles." To see how this applies, first look at the earliest and possibly original form of the bipedal ape as represented by *Australopithecus*: about four to five feet tall, with a chimp-sized brain, and a chimp-style jaw with teeth very much like ours. We know of several species, some one of which led to a split-off or descended form, and some other of which led to another split-off or descended form, in two completely different evolutionary events at different times. Check them out:

- *Paranthropus*: over five feet tall, with a brain proportionately unchanged in size, with an astonishingly increased jaw size and musculature, including a ridge on top of its head, which anchors the chewing muscles. Feel your other chewing muscle, the masseter, the bunchy one at the corner of your jaw—in *Paranthropus*, that fossa (the bone holding it) is about four times larger and the muscle itself is bigger than a golf ball. The surface area of each molar tooth is four times bigger than yours. *Paranthropus*'s skull is very much like a chimp's (or rather *Australopithecus*'s) with the jaw increased in size to account for well over half its mass.
- *Homo*: also over five feet tall, with the cranium expanded in size and the jaw reduced in size, to some extent retaining the typical ape's infantile proportions. I don't need to walk you through its details, but poke your head if you want to, in order to feel the difference. However, remember that our own nigh-absurd cranium size was not a feature of early species of *Homo*, whose crania are definitely distinctively globular in comparison to the others, but not all that much bigger.

Each group was bipedal and had small canine teeth, because each evolved from *Australopithecus* and retained those features from it. In *Paranthropus*, the brain remained unchanged, in relative size at least, and the lower jaw and its support structures became much bigger and stronger, with rather astonishingly large and strong molar teeth, and in *Homo*, the jaw became more delicate and the brain, or specifically the cerebral cortex, became bigger. Beginning with the same template and working parts, different subunits became altered. Significantly, neither is ancestral to the other and cannot be thought of as a way station to the other.

Extend this possibility to most of the distinctive features mentioned in Chapter 5, which aren't preserved in fossil remains. We don't even know when or in what order any of them kicked in: the dense neurological microanatomy, reduced hair cover, aseasonal reproduction, covert fertility, non-ovulatory female proceptivity, protruberant nose cartilage, breasts, increased penis size, astonishing gestation length (including the phenomenon of menstruation), menopause, long life span, dramatic play, language, and inventions or rituals that are taught to later generations. Given the lesson of our skeleton, it's simply wrong to assume either a holistic, collective graded change among only the direct ancestors of *H. sapiens* and no one else, or a complete absence of all of them until the day our species first appeared.

Sole Survivor

In my many teaching experiences about this material, I have learned what the next objection is, always right on cue: But did we not stand while the others fell? Did we not defeat them? Do we not see our triumphant progress out of the dark in the paleontological history? Do we not see the reason why, as we survey that wonderfully increasing head size? Does that not indicate a main branch of ever-smarter human evolution tracked by prototypes and a couple of failed side branches?

It is not a student's fault that our culture is like Prendick and refuses to look. All of these questions are romanticizing what is, for any other group of creatures, a common phenomenon. Is every relict supposed to have a heroic history of selection which granted it a single amazing adaptation that rescued it from the brink of annihilation? There's reason to think that selection was going on in our and the other extant species of *Homo* during the Pleistocene, but that it saved us is not logically indicated—or if one thinks it is, then it's suspiciously absent concerning every other relict. Before treating this idea even provisionally seriously—and then only as a hypothesis requiring rigorous testing—that discrepancy would need to be aired.

Let's stay with *Paranthropus*, who could stand a little general appreciation. It is terribly wrong to characterize this animal or any other extinct member of this group as a confused, foolish, failed group of creatures simply because it's extinct. Not only do we know of exactly zero "stupid" creatures relative to their particular way of life, *Paranthropus* is another great ape, after all, just like the rest of us, not a group noted for dullness. It would be most surprising if these species didn't have a similar degree of socializing and communication as that observed across all creatures of this kind, whatever its particular details were. Yet somehow this "extinct oaf" idea was an unbelievably persistent concept in much academic literature, and if not so much there any more, it still persists in the larger culture. Somehow we're to understand that extinct bipedal apes forgot everything they inherited from their common ancestry with chimps and gorillas, so they tried to communicate, but badly, tried to socialize, but badly, tried to deal with weather and predators, but badly. It's as if in order for our own species' intellectual capacity to be called an achievement, anyone closely related to us must necessarily have been extra stupid, leaving all the achievement to ourselves.

Going by the fossil record, the whole genus persisted for about a million and a half years, from 2.7 to 1.2 million years ago (mya). Of the three currently acknowledged species, which overlap somewhat, *P. aethiopicus* is found for about 0.2 million years (2.7 to 2.5 mya), *P. boisei* is found for 1.1 million years (2.3 to 1.2 mya), and *P. robustus* for about 0.8 million years (2.0 to 1.2 mya). Since the earliest and latest known fossils are not likely to be the very first and very last *Paranthropus* in existence, these are conservative estimates.

The earliest known *Homo sapiens* fossils are 195,000 years old, or 0.195 million years, and estimates based on genetic mutation rates land at about the same time. That means we haven't been around as long as any species of *Paranthropus* persisted—to match the certainly too-short estimate for the briefest known one, we still have five thousand years to go. Before we start talking about what survivors we are, let's try and beat at least a couple of our allegedly stupid close relatives in species longevity, OK?

We may also rightly toss out the self-centered notion of main branch versus side branches, best illustrated by rearranging the boxes any way we like as long as the basic nesting stays the same. All phylogenies are bushes, with no direction or collective terminal point. Being a lucky relict doesn't mean your ancestry beyond your own origin was a "main branch."

It may still strike you that being the single relict of this species group, as well as one of its biggest-brained, must mean *something* special. To evaluate that point, the only relevant comparison can be with those creatures who went extinct during our time on the planet, that is, when we might have gone extinct with them, but didn't. That would be three big species, including *Homo neanderthalensis*, who lived mainly around the Mediterranean Sea, *H. heidelbergensis*, who was a bit more widely distributed in eastern Africa, the Levant, and eastern Europe, and *H. erectus*, who was mainly in eastern Asia; and one small one named *H. floresiensis*, so far known only from a single island in Indonesia.[2] The first two were tall, big-brained, people-looking species, and although we don't know much about their soft-tissue features like nose cartilage, presumably they would be striking to you in life only because their ethnicities would be confusing. The miniaturized species is not too surprising, as that's a common phenomenon in mammalian species assemblages, and even within the single species of modern humans, we include at least five living instances.

You see where this is going, right? *Homo sapiens*' distribution from at least 140,000 years ago onward covered most of Africa and Eurasia, and 15,000 years ago would become truly global, whereas each of these other species lived in a comparatively tiny spot. What you're looking at is the same story for all the land mammal groups at this same time, 110,000 to about 12,000 years ago, somewhat simplistically called the Ice Age: a dramatic rate of extinction for bigger species, for those with specialized diets, for those with smaller ranges, and for those with low birth rates, especially when a species features two or more of these things. That's why today there are only a couple of big cats, a couple of elephants, and so on and so forth, remaining from what used to be very speciose groups of enormous land mammals. Also, of all the continents, the mega-size mammals in Africa, where the majority of *H. sapiens* lived, were hit least hard.

The groundwork and perhaps even the presence of our distinctive heads and cognitive abilities precede evidence of culture and technology; that is, those features did not evolve "for" those things to happen. Human cultural artifacts date reliably only to about 50,000 years ago. All that big-head change happened in the absence of the things we credit ourselves for doing with it.

H. sapiens persisted not because it had some specially adapted trait honed to perfection by which it cheated fate, but because, like every other big mammal that persisted, it did well in those risk variables, in our case a large and widely dispersed population, a fair amount of that population in a less vulnerable region, and a generalist diet; possibly aseasonal reproduction contributed, too, if we had it back then. Our current presence as a relict is evidence of literally nothing regarding our mental performance variables, specifically those that have resulted in recorded culture and in technology. Nor does it imply optimality of any kind, especially not as a circular indicator that the way we are is a designed or refined phenomenon that "beat out" anyone else.[3]

A Technical Problem

It might surprise you to learn that one of the key terms in biology, "species," is not very well defined. For paleontology, the problem is simple: we have scattered pieces, not populations of creatures, so naming a species and arranging their relationships is always going to be provisional at best, and even a single new fossil can prompt big changes in the proposed arrangements. You'll see that in action if you follow up with these or other references; for example, some people consider *Paranthropus* to be in the same group as *Australopithecus* and only use the latter name. But it's far more terrible with living species, as the discipline has not settled on a notion, even just a working idea, of what a species biologically actually is.

This is a long-standing mud-wrestling topic in evolutionary biology, going back to a technical disagreement between Charles Darwin and Alfred Russell Wallace: Are species biologically continuous or intrinsically separate? In Ernst Mayr's epic interpretation of Darwinian theory during the 1940s, 1950s, and 1960s, it seems he solved the problem with his biological species concept, that species are defined by reproductive incompatibility. It's the most cited definition of species in use today. Unfortunately, this concept suffers from the observation that sister species, the ones most closely and immediately related to one another, frequently can and do interbreed in nature, which is to say, it's not true. (For the record, Mayr was right about many thousands of things.)

The vocabulary to discuss this problem descends swiftly into jargon-ridden madness, with adaptive radiation dueling with evolution by entropy, with at least four different meanings for "isolation," and with revised evolutionary arrangements of species sprouting throughout the conversation like fungal fruiting bodies. I dip into it here for one single point, that when a new species evolves, a great many of its features are retained from (or if changed, are subject to constraints from) its parent species. No species that existed was a blank slate to be written upon anew by natural selection.

Selection and Its Discontents

Selection is one of the strongest scientific concepts known. But it's not talked about in similarly strong ways, least of all for humans.

For example, we're products of natural selection, so we're "adapted." That's exciting. Unfortunately this term is long compromised, having served multiple conflicting authors since its introduction. Is adaptation the feature that's been selected for, as in "an" adaptation, or the process of variation and selection that led to it? When a creature shifts among physiological or behavioral states relative to the environment during the course of its life, as in growing fur of different colors in different seasons, is that adapting? Or is the entire suite of features the adaptation, in which case, how many traits is it? Apparently humans are adapted because they're adaptable and live in multiple climates, whereas other creatures are adapted because they're specialized and live only in a very specific ecology. Biologists usually take the specific meaning in context, but they've been known to get tangled up, and the more general teaching and use of the term has descended into near hand waving. I won't even get into the noise surrounding the precise, technical meaning of fitness.

Here's another. Selection is a matter of population averages: it's not "looking for" a single best performance, but the type of performance that most reliably succeeds (this time) becomes more common. This does not lead to a march toward homogenous maximum competence. Even if one variant remains most reliable without fail, if several alternative variants are available, and if the one with the highest success rate clocks in at 40%, then even from the strongest imaginable selection, you'll see a population of creatures who fail when they try it 60% of the time. The popular notion that selection leads to excellence corresponds only to edge cases or reflects our bias toward watching successful and dramatic actions. Behavioral studies in the wild quickly betray this expectation; the most common result is adequacy, and even that adequacy applies only to past reproductive success, not to whatever circumstance one individual might be facing today. Too many variables exist to expect selection to favor a single, outstanding feature as the default outcome.

- One only has so much external information from necessarily limited sources.
- Contingencies of the immediate environment can't be relied upon to cooperate.
- Internal proprioception and coordination of action are subject to contingencies, too.
- Cognitive maps are full of self-generated fibs.
- The movements of others cannot be predicted, only estimated.

It's rather a wonder we succeed at anything at all, as selection only addresses statistical success, and the grotesque or tragic failures of the most selected-for trait can be quite high from a personal point of view.[4]

Many presumed implications about selection need a better airing outside limited academic circles. Here's a quick look at some more of them: the implication that selection perfectly matches a creature to its environment, that an organism's working parts comprise an optimized gestalt, and that selection has brought a creature to a particular state through "tries" in the form of other species. One of the most pernicious is the image of selection as a problem-solver, improving the state of individuals' experience of life or the state of the species in terms of the risk of extinction. However,

- Consider a prey animal's running speed. Has selection resulted in a faster creature, on the average, perhaps much faster? Yes. Has it changed the fact that the majority of individuals die because they're killed and eaten by the same species of predator? No.
- Animals featuring many dramatic selected traits, the very ones that we either share or are impressed by, are more likely to go extinct rather than less. Therefore the term "advantage" needs to be restricted to a very technical meaning, because neither the species as a whole nor the individuals with the selected traits are identifiably better off.

I stress these points because they confirm what Huxley already nailed in the essentials: *natural selection is not about becoming or being happy.* In fact, go ahead and use all the pretty framing of natural selection you want: call us advanced, call us higher, call us perfectly adapted, anything you like—and the fact remains that not one process of organismal change that we know of is oriented toward producing a less-stressed, less-fearful, or to use a difficult phrase, more well-adjusted creature. You want to know the really harsh part? Being a social, cooperative, reciprocal creature makes it *worse.*

What We Do

Many, many animals are extremely social, living in a web of complex relationships and alliances, including most, although not all, primate species. In social species, individuals work together in many different ways, depending on their particular ecology: group defense, group hunting or foraging, group-observed mating rituals, and more.

The most complex version of sociality is called ultrasocial, which has now been studied across dozens of species. Its features are both familiar and, when framed in clinical biological terms, a bit alarming. In an ultrasocial species, the animals live densely packed, and many common behaviors as well as these teamwork activities are effectively collective. These activities are so prevalent that they set the whole circumstances of an individual's life. Individual social roles and behaviors are usually observable to other members of the group and are rapidly communicated to still other members. A key variable for an individual is therefore inclusion in or exclusion from various collective actions and benefits.

Many bees and ants are ultrasocial, with the specification of eusocial—that means the hive or nest members share extremely high genetic identity, and a few individuals' reproductive success literally belongs to them all. One-on-one or individual-versus-group conflicts are rare, even absent, in these species. However, our version of ultrasocial is different: individuals are "alone in the crowd," genetically speaking. Each individual has his or her spoon in the stew of resource acquisition, mating possibilities, and risky cooperation. All sorts of general organismal problems like mate choice, parasite avoidance, and more are ramped up to a whole new quantum of opportunity and risk, not least because whatever one does, a whole lot of others are going to know about it soon. Kin-oriented behavior becomes more intense as well because the genetic relatedness results in a sub-community of its own, itself positioned relative to all the other social groupings.

Therefore in our ultrasociality, and in that of any species with a similar arrangement, complicated guessing and individual conflicts are the norm, because of the larger web of cooperation and expected obligations. Cognitively speaking, it's a huge and shifting puzzle, with family trees snaking through a web of overlapping social groups, every individual point shaded with reputation, constantly modified by events and rumors of events. Ultrasocial animals are really, really good at soap opera because that is, effectively, their most immediate and relevant universe. Sound familiar yet?

Considering how humans have been affected strongly by selection is a reasonable scientific topic, although its meaning for our daily lives has been and to some extent remains a tacit and firewalled topic. A generational shift has occurred such that few would disagree that our current array of behaviors has been affected or even shaped by selection, although it's still typically discussed in a cleanly and rather sterile form, which is also romantic: the shaper of wonders. Taken a bit too far, as I think it often is, we're back in Huxley's early writing, saying that selection made us special, so don't worry, "from apes" or not, we're still special.

I've been professionally frustrated with sociobiology for a long time. At the height of its academic popularity in the early 1990s, it failed to incorporate mammalogy and a general systematic perspective. People studying human behavior in an evolutionary context are too eager to create cool narratives specific to ourselves, in an exact parallel to the century and a half of paleontology. That's why I've been so strict about understanding how much of humanity evolved long before we were a glint in a *Homo ergaster*'s eye, and how much of it cannot be described as "ours" in any serious way. The field seems overdue for a more mosaic, non-optimal, and generally systematic context for discussing human behavior, such that we *are* a mammal and an ape, rather than a special entity who has evolved *from* them.

I'll restrict my discussion to two authors who strike fearlessly into this issue. The first is George C. Williams, specifically his commentary (1989) on *Evolution and Ethics*, in which he states that modern views toward genes, selection, and behavior confirm Huxley's grim portrait of an outright destructive cosmic process. To the list of adjectives describing selection and other ecological aspects of reality, he adds wretched, immoral, unjust, abysmally stupid, and outright evil, and ends at an even more qualified hope, without even the "irrational to doubt" phrase to soften it. Schopenhauer would have loved it.

He explains why: that natural selection is bereft of consequence in terms of individual experiences. If some behavior hurts oneself or anyone else, but qualifies for the mechanical effect of selection, it gets selected for just the same. In social groups, the opportunities for such cruelty and outright horror are legion, and therefore social life is as much a catalogue of atrocity as an effervescent cultural celebration. He rightly points out that mere communication and culture are no solution in themselves, as culturally transmitted information—memes—are just as grim in their effects as genes.

The question is whether reciprocity, or mutual helping, which is technically selfish in genetics terms, nevertheless generates values and practices that mitigate the agony. Such effects would be less powerful than genetic advantage in an all-else-equal face-off, but circumstances are more fluid than that, and there's at least a chance for hope. The idea is that selfishly evolved phenomena have been, and can be, co-opted into more helpful and mutual activities. In this, Williams echoes Huxley's point, just as lonely today as then, that the very worst thing to do with our understanding of evolution is to represent it as a directive or a model or a justification—for oneself, for a proposed social plan, or for anything. It's not there to help us.

Richard Alexander's *The Biology of Moral Systems* (1987) is the strongest text I know about the consequences of ultrasocial life for humans, specifically the condition of cooperating with many, many unrelated individuals, seeing what's "up and past" the behaviors that can be easily identified with direct genetic advantage. I've found that it's a touchpoint for people interested in these topics—"Have you read Alexander?" and if the answer is yes, usually followed by a bunch of ideas and interpretations, then you keep talking to that person.

> ... many moral philosophers do not approach the problem of morality and ethics as if it arose as an effort to resolve conflicts of interests.... as if questions about conflicts of interest arise only because we operate under moral systems, rather than vice versa. (Richard Alexander, *The Biology of Moral Systems*, p. 89)[5]

The question then becomes what kinds of conflicts of interests are faced by an individual in an ultrasocial species.

According to Alexander, the ultrasocial environment plays a huge role in our selective history, to the extent that reciprocity is not only prevalent, but also provides a web of context for all the other kinds of action. Sure, tit-for-tat exchange is present, but that's trivial compared to the complex obligations we impose and receive from a variety of simultaneous communities, themselves replete with individual and collective action. These relationships vary from genetically very connected, as with close kin, to extremely remote, conducted only via communication, hearsay if you will, among third parties. Most of what we're discussing here falls in between, among unrelated individuals who know one another directly or through one or two degrees of separation.

I should take pains to stave off a common misconception of these matters—this is not about competition versus cooperation as absolute qualities, or as opposed foundations for a social system. It's better to talk about individual or collective actions and decisions, combining affiliative and agonistic qualities—that lets us acknowledge group actions hostile to others or toward individuals, defensive violence, and legal punishments as part of the picture. The idea is to identify the active variables in any arrangement of power, regardless of its purported ideology.

What follows is my interpretation of some of Alexander's text and may deviate from or elaborate upon his points in some ways. I think of immediate, personal life concerning people right next to us as "core" problems: most often issues of maturation, romantic connections and commitments, and the range of concerns within family and parenting. The larger community context includes all the circumstances of resource and their acquisition, and the degree of help one can expect and about what. Therefore decisions about all these "core" things are subject to third-party and collective judgments, giving them a second-order set of significance (Figure 7.2). Certain conflicts are associated solely with that larger context, such as the degrees of threat and violence offered toward one another, all sorts of difficulties with group membership and obligations, especially avoiding being exploited, and like a constant haze over everything else, the dance of deception and its detection. Both the second-order form of the core conflicts and the larger-scale contextual ones may be so present and consequential that their impact affects individual fitness (the key variable of selection) more than "basic" foraging and mate choice do.

Group identity and multiple, overlapping communities become the equivalents of weather, resource-rich areas, or disease, providing overwhelming consequences through collective action based on an individual's reputation, accurate or not. Most of these interactions can be understood as direct reciprocity in a dizzying array of currencies and information-gathering, in which a great deal of our employment and commerce resides.

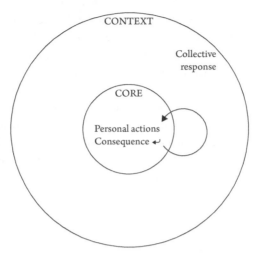

Figure 7.2 Direct and indirect degrees of behavioral investment.

Consider a commonplace but significant personal act, for example, engaging in sexual contact with another person. One might do this in full accord with immediate social expectations, or in doing so, favor one such expectation over another, or do it in defiance of any social expectation. Add in the complications of rehearsing it in one's cognitive map, negotiating about it with others who aren't directly involved, communicating about it during the action or afterward, and finally responding to whatever others' reactions and possibly supportive or retaliatory actions may be. It makes no sense to call the anatomical activity biology and the rest of it something else. This is *all* biology.

Consider actions that are "way out there" in terms of risk and (for lack of a better word) return, without an overt or identifiable exchange. Alexander's text explains that although these actions' immediate benefit is impenetrable to our senses and cognitive predictions in the moment, doing them carries its own weight and possible consequences in the context of this ever-present, often judgmental community. That's where the selection comes from, consistently enough to matter. He observes that such "furthest outward" actions toward one's larger community or toward the helpless are less common and consistent than obvious and immediate fitness-increasing ones, but they are nonetheless present and selected for. He's essentially exploding the idea that motivational selfish concepts are even involved.

An organism whose behavior has evolved in this context would seem to lead a most fraught life, full of opportunities and interferences, labeled with both true and false reputation, bombarded with expectations, trying to reconcile them with both observations and preconceived notions, and constantly trying to adjust relationships and manage one's perceived identity among a widespread group.

One's personal decisions about romance, kin obligations, parenting, violence, cooperating, expecting recompense, and deception would all be under constant judgment—you'd have to figure out whether you're right or they're right, and what you'll do about it, constantly!

We can be ethical, empathetic, decent, forthcoming, and generous because all of these things are within our capacity and have been selected for in an ultrasocial context; and we are simultaneously otherizing, petty, downright mean, bald-faced liars, and capable of atrocity for exactly the same reason. We can do any of these in accord with a collective action, or in defiance of it. All of these options are available on tap, tangled up in the ultrasocial web of unpredictable immediate outcomes and outsized social-level consequences, as managed by imperfect cognitive maps and social identity.

If you'd like to call the experience of coping with all of this, throughout one's life, "moral conflict," I won't argue. One only gets to live once, and each moment happens only once. A person must grab an action from the whole range of possible ones, trading off among investments in others' well-being with one's own, in this huge and potentially disastrous web of community expectations and potential retaliation. One's identity, or rather, identities, both internal and socially perceived, are constantly in question and at stake. This sounds more than stressful; it sounds traumatic. What could a moral conflict even be, except for this?

In using or investigating Alexander's model, the issue is not to discover the single most moral thing to do, or a collective direction or goal for moral actions as opposed to immoral ones, but instead, to discover why we experience specific kinds of external and internal conflict under specific conditions, which we recognize as distinct and characterize as moral. This is not about answering previously insoluble questions; it's about materially explaining why certain things that we already know about are the case. One can abandon the age-old Manichean knots about individual versus societal, internal versus external, selfish versus unselfish, as all of these are simply present without inherent right/ wrong content.

That's why one should neither seek nor fear the moral implications of framing our behavior in biological terms. *No* single option is "the" evolutionary one as opposed to the others, and not a thing about behavior is guaranteed to work in a single instance, whether that means success as visualized in the moment or in terms of biological fitness. All biology can tell us, as for any animal, is the phylogenetic, ecological, and physiological history of how we turned out this way.

Saddled as it is with the Man/Beast divide, our cultural vocabulary is completely inappropriate to express this reality. Discussing social morality has almost always included a prescription for moral action, especially when a description of humans is included. It's a terrible struggle to wrench the conversation away from "what does biology tell us to do" with the limited phrasing available, let alone from "this is how we have freed ourselves from biology," embedded in the biological literature.[6]

Perhaps shorn of our constant chatter about them, our decisions about how to fit in or not fit in, to do well by others or not to, to protect ourselves from being exploited, to work together toward difficult ends, really aren't different from other ultrasocial primates' ordinary decisions. Perhaps, too, we inherited most or all of this without evolving it ourselves, just as we inherited bipedalism. Plenty of cherished "human" actions cannot be claimed strictly as our own, when we observe many aspects of these decisions and conflicts in other ultrasocial species, including sublimated coping mechanisms, dysfunctional stress behaviors, self-sacrifice, and cruelty, as well as collective responses toward them.

The good news is that many biologists agree with this point and have worked hard to establish ties with other disciplines. We've reached the point at which one can find unplanned links among these ideas. It may even be possible to talk about feelings and moral crisis without dismissing nonhumans as robots, descending into sentimentality about their nobility, or striving to find a prescription for virtue. As just one example, Marc Bekoff and Jessica Pierce's *Wild Justice* (2009), Paul Gilbert's *Human Nature and Suffering* (1989), and Sherryl Vint's *Animal Alterity* (2013) create a foundation, including vocabulary by which we might at last ask, what does "human" mean when it's not merely a cherry-picked subset of this particular profile within social primate behavior, but considered as a typical example rather than an atypical one?

Yet again, though, Wells got there first. Despite all Prendick's efforts to describe how awful and distorted the Beast Folk are, the issue isn't how much their nonhuman features disgust him, but instead how he sees and struggles with the *fact* (to use his own word) that real people are looking back at him. What has the Leopard Man done, but a familiar act of petty temptation and rebellion? What has the Puma Woman done, but struck back when she was hurt and exploited?

The Valley

In 1970, Masahiro Mori proposed a useful concept, which he called the Uncanny Valley. He was addressing the issue of adding reassuringly familiar human-like features to a robot or prosthetic device. Adding more and more of them elicits more and more positive responses from people, especially when encountering it for the first time—until you hit the Valley, when they strangely elicit a reversed, negative reaction rather than a positive one (Figure 7.3). It's a retroactive horror, in that you're squicked not so much by the thing, but by realizing it's not human when you were thinking of it as human just a moment before.

I'm applying the concept to Prendick's experiences especially after he knows the Beast Folks' nonhuman origin. Therefore it's not their nonhuman features that

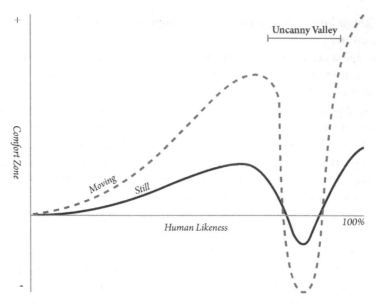

Figure 7.3 The Uncanny Valley.

are distressing, but rather those which are too human, and too many—enough to lull you into false recognition at first. I think it applies quite well. For example, only the sudden appearance of a Beast Man's teeth, when he yawns, reminds Prendick that he's not looking at an ordinary human being—therefore it's not the teeth that freak him out, but rather *everything else*, all of which until that moment *seemed* normal. This and other similar examples place the Beast Folks' repellent qualities more in the Uncanny Valley than in the category of obvious Halloween-animal appurtenances.

Mori's construct isn't supposed to cover every instance of human-like features in all circumstances. It's solely about the tendency for accumulating human-like features to generate a specific mixture of positive *and* negative viewer responses, especially at first contact. Much debate has ensued concerning when the Uncanny Valley does and does not apply, and what is or isn't uncanny. I'm sticking pretty close to his original stated concern: dealing with something discovered to be nonhuman, and whether and when its human-like features make it repellent. Again, apparently this happens only when these features are initially convincing.

Mori also called for a mapping of the Uncanny Valley, which has generated many responses. A Moreau-centered version of such a map would first concern human versus nonhuman animal anatomy, which is well-covered territory for artists, including the nineteenth-century cartoons of Charles le Brun and Thomas Rowlandson, as well as Daniel Lee's modern images (*Manimals, Nightlife*), Kate Clark's sculptures, and Patricia Piccinini's sculptures.

Appearance

What do they look like? The text is unambiguous: anyone who encounters the Beast Folk thinks they are human people, even upon close-up acquaintance. Even those who are repelled by them still think so. They are not, for example, hairy. Nor is their speech distorted or agrammatical. They flatly do not have nonhuman heads atop humanoid bodies, digitigrade legs and feet, or tails, as they are so often portrayed in illustrations. Here is Prendick's most extreme specific description, from Chapter 9:

> ... three grotesque human figures.
>
> One was evidently a female. The other two were men. They were naked save for scarlet swathes of clothing about the middles, and their skins were of a dull pinkish drab colour, such as I had seen in no savages before. They had fat heavy chinless faces, retreating foreheads, and a scant bristly hair upon their heads. Never before had I seen such bestial-looking creatures.
>
> [*The Swine Men begin a dancing, chanting ritual.*]
>
> Suddenly, as I watched their grotesque and unaccountable gesture, I perceived clearly for the first time what it was that had offended me, what had given me the two inconsistent and conflicting impressions of utter strangeness and yet of the strangest familiarity. The three creatures engaged in this mysterious rite were human in shape, and yet human beings with the strangest air about them of some familiar animal. Each of these creatures, despite its human form, its rag of clothing, and the rough humanity of its bodily form, had woven into it, into its movements, into the expression of its countenance, into its whole presence, some now irresistible suggestions of a hog, a swinish taint, the unmistakable mark of the beast.
>
> [*One stumbles and is briefly on all fours*]
>
> ... that transitory gleam of the true animalism of these monsters was enough. (*The Island of Doctor Moreau*, pp. 29–30)

Read in isolation, that would seem unambiguous: look, they're animals. However, this conclusion never occurs to Prendick. All these impressions and his invective ("foul beings") are sound as far as they go, but it is this precise observation that begins his train of thought, concluding that the "savages" are human beings distorted into animals by Moreau. Nor is it a blurry or fleeting impression; he watches the almost-naked Swine Men for a while, suggesting that many telling details—human breasts, for instance, as he implies—do not trigger the "it's a pig" response in him. He is talking solely of how much *like* pigs they are, but never doubts that they're human.

In this map of the Valley, its deepest point could well be identified by Prendick's comments about the "evil-looking boatmen" and other Beast Folk at the beginning of Chapter 6:

> . . . the four men in the launch sprang up with singular awkward gestures . . .
>   . . . this individual [*later identified as the Ape Man*] began to run to and fro on the beach, making the most grotesque movements. (Ibid., p. 18)

As with the Swine Men, the verbal emphasis on how much they're off-model may distract from how much they're *on*. Everything they're doing is tagged as human, behaving completely socially appropriately, working together without being ordered and monitored at every little step, talking casually with one another, or, as in the case of the Swine Men, conducting an ordinary religious ritual of some kind.

In Chapter 15, after he learns of the Beast Folks' origin, Prendick describes a number of their physical peculiarities, including their short thighs, incomplete and clumsy hands, and lack of lumbar curvature, with the net effect that the Beast Folk are pretty awful-looking. However, in the context of that one basic observation, that everyone meeting them does indeed think they're people, then these features can't be so obvious. For him to realize their origins and before he can list those features so fully, he must be forced to witness the surgery, and even his first such witness actually convinces him of the opposite, that Moreau is operating on a human person.

The prose provides many details but never a single portrait, in such a way that my imagination as a reader is thrown into high gear—but in a curious, shifting fashion, not entirely certain of just what the Beast Folk look like, but with a distinctly eerie quality. Not one Beast Person is described with fangs outside its mouth or horns or similar features; the single exception for body shape, the satyr, is acknowledged to have been one of Moreau's fanciful deviations from his project.

In the film adaptations and most illustrations, the Beast Folk are typically hairy, prominently fanged hybrids, especially from the 1970s on, which subverts the explicit content in the book. In *The Island of Doctor Moreau* (1977), their appearance and movements are identical with Lon Chaney Jr.'s performance in *The Wolf Man* (1941).

However, some of the characters and events are more consistent with the Valley concept, especially the various forms of the cat-based woman. In *Island of Lost Souls*, the Beast Folk are notably more human-like in the first half of the story, in many cases consistent with the eeriness of the novel's techniques, but much more hideous in the second half. Also, Ouran (Hans Steinke) in this film is indeed hairy and slouching, but I think he's at his most villainous when performing repugnant human actions, and that his grin is disturbing because it is so unaffectedly appropriate to them. Everyone who plays the Sayer of the Law (Bela Lugosi, Richard Baseheart, Ron Perlman) really tries. The short sequence with the bear-man (David Cass) in the 1977 film gives him surprising dignity and offers the possibility of a good subplot, although the story drops it almost instantly. Most of the illustrations in the Classics Illustrated version are quite monstrous, especially the Swine Men, but the artist Eric Vincent pulled off a remarkable feat with his close-up on the Leopard Man's face, which could serve as the perfect example for the transition in and out of the Valley.

To be fair, visual media will always have a hard row to hoe with this story. At this late date, if people read or see anything titled "Moreau," they're probably going to be disappointed unless they receive a solid monster-fest spectacle, in which case the Uncanny Valley is irrelevant. But even if the filmmakers wanted to present the Beast Folk in this way, they'd have to nail it absolutely perfectly, or the characters would be too recognizable as actors wearing some make-up. Perhaps a careful comparison between Lota (1933) and Aissa (1996) can be made by someone more in tune with film techniques than I am, because I cannot quite explain why the former character seems visually successful to me and the latter does not.

That observation calls attention to what lies at the "near" end of the Valley, though, at its upswing. For this is what matters most, when the "off" features are at their most minimal and the startling, even terrifying effect occurs all the more unexpectedly. At that point, it's the on-model effect that matters most: the things about this creature that are not scattered details but the entire context, and not only human-like, but unambiguously so, not merely convincing but actually and genuinely identifiable as "ours." At that point, the shock has nothing to do with "oh my, it has fangs, it might bite me." It comes from invoking too much self-reflection. There I think one unpleasantly finds that we ourselves are not out of the Valley when scrutinized too closely, as well as the hint that we rely too much on comforting appearances when assessing someone or something for humanity.[7]

Prendick soon discovers this for himself, regarding the Beast Folk's undeniable sameness with ourselves in two activities well known to our species that have so often in human history been tagged with, as he calls it, the unmistakable mark of the beast.

Sexuality and Women

I was surprised to realize, upon closer reading of what appears to be an all-men all-the-time story, that women are present and significant throughout most of it, and that they are distinctly thinking and behaving as people.

The Beast Folk do marry, although no relationships, romance, or rituals are depicted or developed.

Beast Women "display a more than human regard for the decencies and decorum of external costume," as he presumes out of their awareness of their "own repulsive clumsiness," revealing that these women are capable of self-esteem and the desire for attractiveness in their understanding, values, and responses.

I may be reaching, but it seems to me as if there's more here than Prendick is saying. No dialogue with any Beast Woman is included in his account, nor any scene explicitly describing their actions as individuals, aside from the Puma Woman. He apparently avoids the Vixen-Bear, whom he dislikes. He describes nothing of Beast Folk romance, of their marriages and daily relationships, but their sexuality hangs over some passages like a cloud.

They are subject to doctrinal morality: the Law is explicit regarding sexual privacy and monogamy, and my vulgar imagination suggests that the parts Prendick specifically censors from his readers (" . . . to the prohibition of what I thought then were the maddest, most impossible, and most indecent things one could well imagine" [p. 42]) concern the physical positions of copulation as well. *Not to Couple from behind;* **that** *is the Law. Are we not Men?*

The women of the Beast Folk are vulnerable to sexual harassment and assault: ". . . subject to much furtive persecution contravening the Law regarding monogamy" (*The Island of Doctor Moreau,* p. 62).

During the events in Chapter 21, the Beast Women also revert more quickly than the men, deliberately disregarding "the injunction of decency," and attempting "public outrages on the institution of monogamy." I infer from these euphemisms that the island becomes quite the female-led free-love den of iniquity for a while, likely contributing to Prendick's earlier statement about this period that "[t]here is much that sticks in my memory that I could write, things that I would cheerfully give my right hand to forget" (Ibid., p. 95).

All of these points seem consistent with mainstream Victorian notions about female human emotions: that they are strictly culturally controlled bestiality just below the surface. In other words, this text confirms the genuinely human qualities of the Beast Women, as expressed by an 1890s writer to his contemporary readers.

It's not all historically specific stereotype, however. One of the few details that slips through the silence about the six weeks may be a crucial moment for Prendick in his entire adventure, when he almost admits that he was attracted to a female Beast Person, only put off at the last second.

> Or in some narrow pathway, glancing with a transitory daring into the eyes of some lithe, white-swathed female figure, I would suddenly see with a spasmodic revulsion that they had slit-like pupils, or, glancing down, note the curving nail with which she held the shapeless wrap about her. (Ibid., p. 64)

Although his very next phrase references the Beast Women's "repulsive clumsiness" in general, I see no evidence of it when it comes to this specific example—rather the opposite, in fact. And wait—daring? Daring to what? I can't possibly be the only reader to infer that Prendick very nearly, almost, just about propositioned a Beast Woman. "Much furtive persecution contravening the Law regarding monogamy," indeed! What did Prendick almost get up to, *or actually do*, during those six weeks he fails to include in his account?

Alcohol

Prendick describes himself with no small snottiness as an abstainer from birth, which deserves a little historical attention. The campaign against alcoholic beverages has a long, involved history throughout the nineteenth-century United Kingdom, briefly described as a series of attempts at legislation, some of which were successful and long lasting. After the founding of the original Salvation Army in 1864 and the politically effective National Temperance Foundation in 1884, the movement was no longer fringe, and was widely considered unquestionably constructive in activist circles. Most other causes included temperance wings, and after it jumped the Atlantic, it would grow in strength to the level of US national policy by the 1920s. By the time of the novel's setting in the 1890s, an adult British person could well have been brought up by a mainstream activist family determined that their children should never be tainted by "drink." What does this mean in late nineteenth-century terms?

Prendick is culturally associated with, even indoctrinated with, the ideals of a movement whose members perceived themselves as bettering society and indeed humanity through social engineering. Temperance activists saw themselves as liberated moderns riding a cresting, climactic wave of unique political action, freeing the less fortunate of humanity from booze as a force of oppression and self-subjugation—in fact, in the language of the time, "making a beast of oneself," another excellent example of using exceptionalist language to "otherize" a human activity. Some of them were dedicated religious reformers, others were atheists, others socialists, and the boundaries among these were hard to discern.

Prendick's teetotalism is his marker, even a dead giveaway, for his own version of Man, which he thinks mere humans should acknowledge and aspire toward—and just as toxic to his own decisions and perceptions as Moreau's never-defined rationality was for him.

Montgomery is equally placed in history, as a former and apparently not very good medical student disgraced by some unseemly act. The Royal College medical students were not well liked during this period, considered ruffians both for their association with vivisection and for their rowdiness, with their violent role in the Brown Dog affair not too far in the future. Apparently they enjoyed some freedom from prosecution, which suggests that whatever Montgomery was mixed up in was pretty bad. In Chapter 8, while the puma is screaming in agony, he mentions "the cats of Gower street," which could mean the literal animals used in medical demonstrations or perhaps the prostitutes, whom it's impossible to imagine he did not frequent. Coral Lansbury's *The Old Brown Dog* clearly explains how the two issues might be combined in the politics and culture of the day, and Montgomery seems like he would be overwhelmed by both. He had asked Prendick about the Caplatzi, likely a whorehouse, and later Moreau refers to his objections to the work during the early days of their arrival at the island.

Unlike Prendick, Montgomery knows what is going to happen to the puma, specifically to her voice. His heavy drinking to drown out her cries is no longer partying; it's self-medication, and probably has been for the better part of eleven years.[8]

Prendick and Montgomery therefore fall squarely on either side of the alcohol issue, reinforcing all sorts of differences they perceive about one another, which had led them to needle one another regularly about it almost as soon as they'd met, especially in Chapter 8. It comes to a head after Moreau's death, when it's no wonder that fresh from killing all the animals restrained in the surgery with Prendick, Montgomery wants to get hammered, and he calls for M'Ling and then other Beast Folk to join in. Prendick initially tries to stop him with a lecture, and the argument culminates with Montgomery brandishing his pistol. During the exchange, Montgomery calls Prendick "you logic-chopping, chalky-faced saint of an atheist," reflecting the view that the ostensibly non-churchy social reformers are no more than Puritans back in action, and says of M'Ling, "He takes his drink like a Christian." Stung, Prendick retorts using the precise language one would expect: "You've made a beast of yourself, to the beasts you may go."

The films maintain Montgomery's combination of bad conscience, substance abuse, and some element of moral failure. In *Island of Lost Souls* (1932) Montgomery (Arthur Hohl) is closest, but this film and *The Island of Doctor Moreau* (1977) standardize him as the minion who finally snaps from resentment of Moreau, in the first helping the newcomers to the island to escape, and in the second (also named Montgomery, played by Nigel Davenport) meeting his end, as many such minions do, gratuitously executed by his former master.

In the others, he's villainized in a variety of spectacular and sometimes cheap ways, including Walter Perrera (Oscar Kesse), the perverted brute in *Terror is a Man* (1959); Steinman (Jan Merlin), the stereotyped homosexual-sadist and a mercenary killer to boot in *The Twilight People* (1973); Montgomery (Val Kilmer), the drug-addled, also sexually ambiguous psychopath in *The Island of Doctor Moreau* (1996); and perhaps oddest of all, Pak (Ling Aum), the hatred- and guilt-ridden father who lurks about with what seems to be a huge scythe and eventually kills Moreau before his own end arrives.

Only in *The Twilight People* (1972) do the two characters have a similar, juxtaposed, train-wreck relationship, based on the differences in their expressions of male power, with Farrell developing his romance with Dr. Gordon's daughter, and Stern crushed out on Farrell.

Why does Prendick object to Montgomery sharing booze with M'Ling so strongly that he only desists when Montgomery holds him at gunpoint? It interests me that Prendick cares so much—he had not reacted so strongly even when seeing the animals restrained and mutilated in agony. If the Beast Man is really "that beast," as Prendick puts it, then so what? Why bother trying to stop a mere beast from being corrupted? Their dialogue in this scene struggles in the grip of exceptionalism, with Montgomery insisting that M'Ling is a real or whole person and therefore competent to enjoy a nip (as well as telling Prendick "*You're* the beast"), with Prendick also identifying M'Ling as human, specifically a less capable human from whom liquor should be withheld. They keep talking about animals and beastliness, but the whole discussion has nothing to do with M'Ling's nonhuman origins, but only with what kind of person he is.

I hope you can see what's present by its very absence: why don't they ask M'Ling himself, who is standing right there and perfectly capable of speech? What is M'Ling's story, anyway? It's easy to miss because he's so quiet, but I think this very silence matters a lot. He's the Beast Man who buys into what the people around him want and gives it a good try.

Montgomery treats M'Ling exactly as a contemporary Englishman might treat any colonial servant, with favors, patronizing humor, and occasional abuse, but always including him in his (Montgomery's) ordinary activities. And insult or pummel M'Ling as he might when drunk, he never whips or harms him, and he also defends him at some risk to himself on the *Ipecacuanha*. However, whether Montgomery is so desperately lonely that he talks a bit too much to his dog, or genuinely sees M'Ling as a fellow person, or something in-between, is irrelevant. What does M'Ling think?

> It was a complex trophy of Moreau's horrible skill, a bear, tainted with dog and ox, and one of the most elaborately made of all the creatures. It treated Montgomery with a strange tenderness and devotion; sometimes he would notice it, pat it, call it half-mocking, half-jocular names, and so make it caper with extraordinary delight; sometimes he would ill-treat it, especially after he had been at the whiskey, kicking it, beating it, pelting

it with stones or lighted fuses. But whether he treated it well or ill, it loved nothing so much as to be near him. (*The Island of Doctor Moreau*, p. 63)

It's hard to say. He's definitely subordinated, treated like a half-person, in the only detail of the novel that explicitly invokes colonial imagery. Maybe he's just a door-mat kind of guy, the stereotyped rather pathetic dog (in *Island of Lost Souls*, he is in fact made from a dog). The question is whether he chooses to accord with this role, understanding it to be the only "real human being" role available to him, and possibly even to value it and Montgomery himself.

Positing that he might hold a relevant opinion, why does he tacitly participate in this argument in the role of an object? I think I can see it. Socially, he's screwed. In the context of their very precise and defined argument, if he drinks with Montgomery who badly needs a friend, he becomes a person, "like a Christian," but if he refuses, effectively siding with Prendick, he only gets to be a less-than-whole person or "that beast."

His silence defeats the more subtle or motivational side of these questions. Maybe none of it applies, and he's merely an insensate brute, as Prendick persists in calling the Beast People, especially at this point in the story. But I don't think so—because when it comes to actions, whether enduring abuse, killing other Beast Folk with his teeth, and now when argued over like a piece of morally incompetent meat, he never abandons his friend.

In this case, that decision (if it is one) is tragic, as the revelry culminates in a horrific, lethal melee. The story's overall portrait regarding Beast Folk and alcohol is both depressing and accurately human: they party like people, they brawl while drunk like people, and they kill one another when brawling drunk if they have weaponry available, like people.

Where we stand

Everyone likes to fiddle with the Uncanny Valley model a little, and here's my idea. One should ask most sternly, what is out of the valley, at our presumed end? Maybe we aren't actually there at all. Maybe when we really look at someone without painting it over with Man/Beast terms and concepts, he or she is ever so slightly, but definitely, not "here" in the safe end, but ankle-deep in the near end of it. Maybe our capacity to make perfectly ordinary humans into the Other has a lot to do with evicting them from our illusions—still comfortably applying them to ourselves—and into the Valley. And if so, that would also explain why bringing nonhumans into our sphere of trust or community often entails shrouding them with the same illusions.

That may be why the Beast Folk are so harsh to contemplate, and the impact of the Uncanny Valley is strong. Not because they are monstrosities, but because they so clearly trigger the recognition that we are ourselves not out of the Valley, because there *is* no "human end" that any depiction or even any real human person could attain, not without the observer's complicity in blinding themselves. Montgomery drinks to keep from seeing the puma this way; Prendick similarly distances himself from her with the odd hatred that emerges somehow during those six weeks and is closely tied to his sugary fantasies about the perfection of humanity back home.

No one wants to talk about seeing ourselves this way, as is painfully evident in the discourse of our own evolutionary history, and even throughout the rest of biology. The discipline has much to answer for in this regard. Even in its technical narrations and depictions, we embrace tactical omissions and distortions to keep from seeing ourselves in that Valley, preferring a rather pretty but incomplete portrait of "achieved humanity," "conscious," "smart," as designed, improved, and perfected by some special, otherwise never-before-seen kind of selection. The human-specific vocabulary associated with medicine maintains an extensive list of synonymous terms separating us from creatures with exactly the same parts, like the appendix instead of the cecum, the thumb instead of the hallux, the stapes instead of the hyomandibula. The most famous reconstruction of Lucy (*Australopithecus afarensis*) shows her trotting along, cleverly avoiding her standing there in full postural homology to ourselves; others often include an individual inexplicably grasping a nearby branch as if it needed a little moral or mental support with its feet so "newly" on the ground. *Paranthropus* appears in no funny memes, is in fact absent from popular culture, because it's off-message, failing to play its subordinated role as a prototype. Our depictions of other species of *Homo* slump their shoulders, cant their necks forward, and thrust out their jaws, despite no identifiable difference in the anatomy that would denote such a posture.

Look at our preferred depictions of our own early species history, because it's careful not to be too early: we're carrying atlatls and bows, poking at fires, building and living in shelters, painting murals, using tools to grind grain, always after that point that begins only the latter third of our entire species history. There is not one portrait of *Homo sapiens* prior to about 50,000 years ago, looking just like us, being ourselves, but in the absence of these technologies and other trappings of accumulated culture.

It's the last Man/Beast taboo, after all—our culture. What if we lived "like animals" for fifty thousand years, even with all of our vaunted biology in place? What if other creatures' social lives were full of deeply felt judgments and confusions, so they are not grunting, huddling, accidentally cooperating, or doing something interesting only when suddenly wrenched by some programmed instinct? What if our moral sense is the way a mammal feels in our kind of socializing, rather than "highest thought?" Previous centuries' discussion of humans in evolutionary and taxonomic terms challenged the "higher human species"— this one challenges the "higher human society" and encounters outrage of its

own. It's all the same: refusing to see the physicality, refusing to see the com-
monality of behavior, demonizing and caricaturing what nonhumans do, and
cognitive tricks to deny it all.

I'm talking about looking at ourselves as humans with a very different meaning to that phrase. We don't say "nonwolf animal" when talking about any animal
who's not a wolf. Nor when we say "wolf," do we mean a ton of things that every
other animal is automatically shut out of possibly having, even its own differing
version of them. Once you remove that special status from "human" as a term,
saying "human animal" would be redundant, like "wolf animal," and saying
"nonhuman animal" wouldn't make any sense, prompting instead the puzzled
inquiry as to which animal you might be talking about. In Linnaean terms, the
only meaning remaining for "human" would be a *common name* or, amusingly,
vulgar epithet. That might seem like a comedown. But what, after all, is so bad
about it?

Readings

See Stephen Raup's *Extinction: Bad Genes or Bad Luck?* (1992) (extinction is not
an endorsement of the survivors—a departure from Darwin and from most of the
twentieth century); John Lawton and Robert May, *Extinction Rates* (1995); and
Robert Martin, *Missing Links* (2003).

Roger Lewin, *Bones of Contention* (1987), applies the points from Misia
Landau's dissertation, which would later become *Narratives of Human Evolution*
(1991), to debates in human paleontology. The narrative in question has shifted
subtly from how we uniquely got here, now that we can see it wasn't unique, to
why we're the one here now. Use your own judgment about how much these ideas
apply to Lewin's own textbook on human origins. See also Ian Tattersall, *Masters
of Our Planet* (2012); Chris Stringer, *Lone Survivors* (2012); and the more popularized text, Chip Walter, *Last Ape Standing* (2014).

The impact of the Piltdown hoax on technical human evolutionary theory,
the political-religious rejection of evolution in the United States, and on the
popular imagery of human origins is immense. The library about it includes
the first publication after it was finally discovered, by J. S. Weiner, *The Piltdown
Forgery* (1955), then Charles Blinderman, *The Piltdown Inquest* (1986); Frank
Spencer, *Piltdown* (1990) and *The Piltdown Papers* (1990); and Miles Russell,
Piltdown Man (2004).

The evolution of the extra-enlarged cranium in the most recent human or
related species is still under debate; a good recent account is Rob DeSalle and
Ian Tattersall, *The Brain: Big Bangs, Behaviors, and Beliefs* (2014). In "Evolution
of the Brain and Intelligence," *Trends in Cognitive Sciences* 9(5): 250–257, 2005,
Gerhard Roth and Ursula Dicke suggest that density and interaction of neurons

is a primary variable, with brain size being less of a deal. As applied to human evolution, the question is whether those neural effects evolved before the brain size increase.

The topic of speciation could occupy us for years. For the technically minded, Ernst Mayr's seminal works include *Systematics and the Origin of Species* (1942) and *Animal Species and Evolution* (1963), and a somewhat different, provocative model is found in Daniel Brooks, *Evolution as Entropy* (1988). The problem of species concepts is summarized in one of the single best review papers in evolutionary biology, B. D. Mishler and M. J. Donoghue, "Species Concepts: A Case for Pluralism," *Systematic Zoology* 30(4): 491–503, 1982.

The primary texts of sociobiology include John Maynard Smith, J. B. S. Haldane, and Robert Trivers, *Natural Selection and Social Theory* (2002), and Edward O. Wilson, *Sociobiology* (1975). The latter is a most contentious topic, so one may as well read it directly before being indoctrinated about what it says. Other useful points are found in Timothy Goldsmith, *The Biological Roots of Human Nature* (1994); and Matt Ridley, *The Origins of Virtue* (1996). Before digging into any of these, I recommend George C. Williams's analysis of *Evolution and Ethics* (1988) as included in the Princeton University Press edition. He provides useful help with some technical evolutionary terms, including the crucial distinction between mechanistic and reductionist, the point that the phrase "gene for" does not imply a single gene or the absence of conditional-developmental effects, and that most genetic change due to selection is a shifting of current frequency rather than full replacement of one thing by another. He also addresses an ongoing problem with discussing behavior, that phrases like "want" and "ought" mean very different things in game theory and sociobiological discussion than in ordinary conversation.

I recommend reading work of this kind like the spokes of a wheel—no single piece being "it," but each moving toward the others. It pays off especially well with Richard Alexander, *Darwinism and Human Affairs* (1979) and *The Biology of Moral Systems* (1987); Paul Gilbert, *Human Nature and Suffering* (1989); Peter Singer, *The Expanding Circle* (1981); Robert Axelrod, *The Evolution of Cooperation* (2006); and Frans de Waal, *Good Natured* (1996) and *Primates and Philosophers* (2009).

It's also frustrating that an evolutionary perspective on humans has fallen into a simplistic Left/Right construction, in which the former rejects the notion that humans are not perfectible, therefore leaving the field to neoliberal and neoconservative co-optation, both of which correspond to the traditional Right. Peter Singer writes about this issue with great honesty in his brief but essential *A Darwinian Left* (1991).

Marc Bekoff and Jessica Pierce, *Wild Justice* (2009), presents the point of view I share and have most often encountered among biologists, that social animals experience moral conflict and what we call virtue without need for special explanations in either direction.

Notes

1. The definition and membership of *Homo erectus* is an ongoing headache, and I am restricting my inclusion of the term to the narrowest interpretation, as an East Asian species. The identity of some African fossils from the relevant period, formerly assigned to that name, is currently undergoing review.

2. "Competition" is another term that has served too many masters during the past century and a half. It's been applied to different species' survival and extinction, to conflicts over range and resource use, and to individual social dominance, in addition to its general meaning concerning genetic variations within a species—all of which have been marred by making inferences about recommended human policies.

3. The popular Darwin Awards summarized in the books of that title by Wendy Northcutt are completely misnamed. Selection does not operate as a judgment on individual mistakes, and although I get it as a joke, the usage still promotes the incorrect idea that selection improves a species in some way.

4. Politics and Darwinian theory have a bad history, mostly at the latter's expense. I am shocked to see even today that evolutionary terms are purported to justify specific policies, gender interactions, or so-called cultural "advances." One culprit seems to be the difficult word "selfishness," which is both loaded with emotion and impossible to assess, as it involves intention. In *A Darwinist Left*, Peter Singer provides many useful thoughts about the rejection of Darwinian thought by a significant portion of the political Left, rightly tagging the good old ideal of the perfectible (i.e., no longer Beastly) Man. It's especially galling that sociobiological terms were therefore available to be co-opted by the political Right, in a grim recapitulation of the same phenomenon just over a century ago.

5. Richard Alexander, *The Biology of Moral Systems*, published by Aldine de Gruyter, 1987.

6. As I see it, the horror of *The Stepford Wives* (the 1972 novel by Ira Levin and the 1975 film) is not that the fictional women are robots, but that actual men saddle their wives with robotic expectations, and when given the opportunity, are happy to replace the annoyingly human women with preferably compliant robots. It's not about deceptive robots being inhuman, it's about actual men who aren't as human as they ought to be—and yet are all too familiar in exactly that feature.

7. I speculate that Montgomery briefly appears in the opening paragraphs of Wells's "A Slip under the Microscope," published the same year as *The Island of Doctor Moreau*.

8. Recent philosophical work is breaking new ground in Man/Beast discussions; see Nicole Anderson, *Cultural Theory in Everyday Practice* (2004), and Gerard Kuperus (editor with Marjolein Oele), *Ontology of Nature* (forthcoming).

PART IV

NO ESCAPE

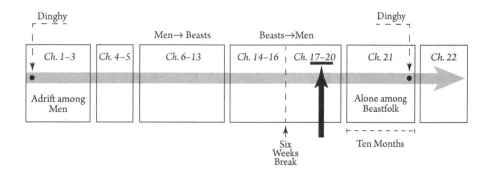

For the world is Hell, and men are on the one hand the tormented
souls and on the other the devils in it.
> —Arthur Schopenhauer, "On the Sufferings
> of the World," in *Parerga and Paralipomena*,[1] p. 48

We enjoy a safe distance from the intensity of nineteenth-century debates.
Who cares whether Empire Brits got their panties in a wad over "man came
from monkeys," or "vivifying forces?" The same can't be said, though, for
the same attention turned toward *our* social constructs: the presumed foundations of our laws, the sense of right and wrong that comes from convention and from struggling with convention, and how all this relates to one's
personal experience, emerging convictions, and sense of agency. Placing
these in the context of human-as-animal sets off modern professional

intellectuals such that poor old Samuel Wilberforce looks like a paragon of restraint, or worse, brings new Spencers out of the woodwork.

As with all the best taboo topics, the meat-and-bones of human society is functionally invisible. The real questions are never asked. What are societal values, religious or otherwise, even for? Does crime defy or disorder a community, or is it part of it? Is there actually such a thing as "an economy?" Is there such a thing as "high" and "low" social conduct? More harshly, how come whatever we do never works as advertised?

Intelligence is not the key. Human symbolic or logical construction is supposed to encourage better human action, whether individual or collective, but it may be that assigning meaning to things and being moved by it does not itself have much meaning. Morality is not the key. Human moral action is supposed to have access to the genuinely cosmic and practical right thing to do, but it may be that experiencing and muddling through moral situations does not yield or reveal a higher good.

Finally, then, what sort of social animal are we? An ideal or maximally functional social order is supposed to exist, which we can find if we think hard enough or try hard enough or educate hard enough or kill enough of the people we don't like, but it may be that organizing social institutions does not result in, or even has nothing to do with, a systemic or advanced society. We may not even have a civilization distinct from mammalian socializing, merely a technological elaboration of it.

No wonder Prendick comes to hate the Beast Folk so much. Their existence commits the unforgivable act of asking and partly answering these questions.

Note

1. Arthur Schopenhauer, "On the Sufferings of the World," in *Parerga and Paralipomena*, 1851, published as *Essays and Aphorisms*, edited by R. J. Hollingdale, Penguin Books, 1970.

8

That Is the Law

The Beast Folk are religious, which, as Prendick keenly perceives in Chapter 16, aggravates their crisis. The issue takes on new shape, however, after the six-week hiatus in his account and after Moreau is killed, and I think these events bring the Beast Folk even more into the recognizable experience of real people.

Regardless of whether it means anything in a spiritual sense, religious behavior is widely observed among humans, and as far as communities go, is universal. Even in the absence of spiritual musings, it plays a major role in organizing difficult social choices. Either we can't recognize it in nonhumans or they don't do it, but either way, religious behavior has been included in the mythology of exceptionalism, as usual with circular logic, that we are special because we do it, and we do it because we are special. This is in addition to the metaphysical claim found in all religions that humans have favored status regarding contact with divine or spiritual forces, and that such forces take a special interest in humans as opposed to other living creatures.

Textually, the Beast People are religious in every positive and negative sense of the word, including the writer-contemporary narrative of Christian history. This is not to argue that they are in contact with the divine or are otherwise spiritually elevated, but rather, that whatever religion is in terms of concrete behavior, they do it fully. It's part of my argument that they are not stumbling caricatures of humans, but that Moreau has succeeded in establishing human-specific features. One may be distressed at the religious qualities of the Beast People, including the ease with which the religious community may be turned against a scapegoat and the readiness of anyone powerful to co-opt religious language into authority, but these qualities are not bestial versions of incomplete or programmed religion, but rather are the same as ordinary humans' behavior.

Like real people, the Beast Folk struggle with their intellectual, social, and ethical choices in mainly religious language, and at the very least, when examining the inner life of such characters as the Ape Man or the Hyena-Swine, it's important to understand that language—biologically, generally, and specific to the period of the novel.

Here my topic is religion as organized social behavior, which doesn't have much to do with religiosity, or belief in the most literal sense. The discussion is made difficult by the word "belief," which may stand for one's observance and sense of community identity without reference to metaphysics. My perspective is that religiosity is not the source of moral intuition, but that moral stress and its resolution are typically framed in religious language and are dramatized with metaphysical imagery. In this context, real creatures live among one another and observe religious practices as part of sharing community, for good, ill, or indifferent, and religious institutions represent a considerable exercise of social power, especially education. One can therefore discuss this "real thing" religion without any involvement of metaphysical matters or, if you prefer, with no concern about God.[1]

To help navigate this murky intellectual landscape, here are some quick definitions:

- *Religiosity* is the personal experience of spiritual connectedness to a cosmic or extra-physical storyline, with strong motivational content.
- *Observance* is composed of rituals, routines, and habits, which reinforce social identity by demonstrating membership in a community.
- *Institution* ("church" or equivalent) is a social organization with political and economic power, which exerts teaching influence and enforcement over people's roles and obligations.
  - *Doctrine* is the written content of such an institution, especially in teaching.

These are three very different things, so I'm going to be strict about which one I'm talking about at any given moment. For example, "the Law" can mean doctrine, meaning the Law's textual list; observance, meaning the Beast Folks' own rituals like the recitation of doctrine, and the funerals; or the institution of power and punishment, which, although it's conducted mainly among themselves, is sometimes reinforced or exploited by Moreau.

Whose Law?

Where does the Law come from? I'll start with where it doesn't. According to Montgomery, Moreau has brainwashed them, to use the modern term:

> In spite of their increased intelligence and the tendency of their animal instincts to reawaken, they had certain Fixed Ideas implanted by Moreau in their minds, which absolutely bounded their imaginations. They were really hypnotized, had been told certain things were impossible, and certain things were not to be done, and these prohibitions were woven into the texture of their minds beyond any possibility of disobedience or dispute. (*The Island of Doctor Moreau*, pp. 60–61)

Well, that's what Montgomery tells Prendick, anyway. I can't find a single instance of expression of this phenomenon in any act of a Beast Person throughout the story. Prendick states this point as part of the justification for why the Beast Folk were not automatically attacking one another or their creators, which is patently something some of the Beast Folk *do* feel the urge to do, so the only possible conclusion is that the techniques are nowhere near as effective as Montgomery claims.

Furthermore, the text is also clear that the Law isn't the same thing as these allegedly conditioned prohibitions. The Law concerns standards by which the Beast Folk try to suppress various urges that they do constantly feel and can indeed act upon, so the actions in question obviously aren't "beyond any possibility of disobedience or dispute." Where did it come from? Again, not from Moreau. Regarding the Beast Folks' lives and activities, he says:

> It's [*Montgomery's*] business, not mine. They only sicken me with a sense of failure. I take no interest in them. I fancy they follow in the lines the Kanaka missionary marked out, and have a kind of mockery of a rational life—poor beasts! There's something they call the Law. Sing songs about "All Thine." (Ibid., p. 59)

The Beast Folk apparently constructed the Law themselves: they were influenced by Moreau's conditioning (such as it may be) and the missionary's teachings, but their spoken Law is their own articulation and ritualization of those teachings. When Moreau invokes the Law to the gathering in order to expose the Leopard Man—whom he does call "the sinner"—he shows that he and Montgomery have learned to exploit the Beast Folks' own culture, but they did not impose it in the first place. I even speculate that Moreau and Montgomery may not be its primary enforcers: the burn scar on the Ape-Man's hand seems quite crude compared to Moreau's methods.[2]

The Law has nothing to do with serving Moreau, either. There is no "fetch and carry for the Master of the House of Pain" in it. The colonial house-servant imagery is strictly from the films; in the book, only Montgomery has trained or enlisted a single Beast Person to his service, which doesn't include Moreau or the functions of the household.

The next issue is whether it's genuine. One could, I suppose, write off their religion as mere imitation and prattle. However, nothing textual supports that interpretation. The quick but precise portraits of many Beast Folk in the book are especially familiar concerning their personal takes on the Law. The Satyr is the classic skeptic, appropriately enough. The Sayer of the Law is the prelate, and going by the scattered descriptions of his appearance, I speculate that his animal origins are both sheepdog and sheep. The Ape Man is the religious intellectual who loves to engage in doctrinal discussion but stays firmly in the fold, especially when he relishes the impending punishment of the Leopard Man. We don't get to see nearly enough of the Vixen-Bear "old woman," "said to be a passionate votary of the Law," that is, the moralist-fanatic, whom progressive Prendick understandably loathes. I'd like to know what she made of his "Big Thinks."

Taking one's religion seriously is more complicated and interesting than merely observing routines perfectly or obeying institutional decrees without dissent. Prendick's description of the Law as something they are "ever repeating and ever breaking" speaks volumes to me—is that not the essence of doctrine, in practice? It's clinched with the funny bit in Chapter 16 when the final Beast Man eventually arrives at the convocation:

> The earlier animals, hot and weary with their groveling, shot vicious glances at him. (Ibid., p. 69)

Although Prendick calls them "animals" here, their mental processing distinguishes between religiosity and observance. The Beast Folk are neither hypnotized robots nor puzzled brutes, but experience and express their religion exactly as real people do.

The films that employ this element always make Moreau into the Law's architect and depict the Beast Folk as much more confused within its strictures, to the extent that one can't be sure that they experience it as anything except operant conditioning. There are a few exceptions and elaborations, though. The 1977 and 1996 films each feature a Beast Folk cremation; the first is accompanied by a funeral march and has a certain dignity, but the second is a dismissive disposal of trash, contrary to Moreau's orders.

In all three mainstream films, the Sayer of the Law is a local enforcer of Moreau's will, but also a true believer, most indignant when Moreau's insincerity is exposed, in a fifty-fifty mix of dignity and torment.

In *Island of Lost Souls* (1932), Ouran takes the Law quite seriously and is shocked when Moreau himself says to contravene it; when Ouran reports this to the Sayer of Law, the latter leads the revolt against Moreau, voicing the Beast Folks' self-recognition as abominations.

In *The Island of Doctor Moreau* (1977), Moreau is particularly vicious and authoritarian, colder than the 1932 version but no less cruel, and the Sayer rather pathetically buys into the power structure—visibly more kempt than the other Beast Men, constantly but ineffectively pleading for them to be "men." When they witness Moreau kill Montgomery, and when they recover the body to examine it, the Sayer not only leads the attack on Moreau, he—unlike the character in *Island of Lost Souls*—stands up to Moreau with some self-assurance, confirming his sense of self-worth by conforming to moral standards and finding Moreau wanting.

In *The Island of Doctor Moreau* (1996), Moreau deliberately acts the Pope, identifying his role as "Father" in both creative and religious terms; furthermore, the Beast Folks' frequent medication makes "opiate of the people" literal. At least some of the Beast Folk recognize this as oppression; insofar as the garbled story can be said to have a moral voice, the Sayer provides some piece of it: "To go on two legs is hard. Perhaps four is better anyway."

The issue at hand is the method of exercising a religious institution's social power. To paraphrase Alexander's points, the two levels of ultrasociality I described in Chapter 7 can be each divided for a four-part diagram (Figure 8.1). Think of each ring as its own sphere or zone of activity, defined by variables of relatedness and direct social association.

- Within the heavy dividing line are interactions that include genetic relatedness
  - Mating partners, children
  - Non-reproductive kin, including uncles and aunts, cousins, and one's parents
- The outer zones concern social cooperation or association among unrelated persons
  - Explicit or near, with identifiable exchanges and community identity
  - Few or no apparent connections.

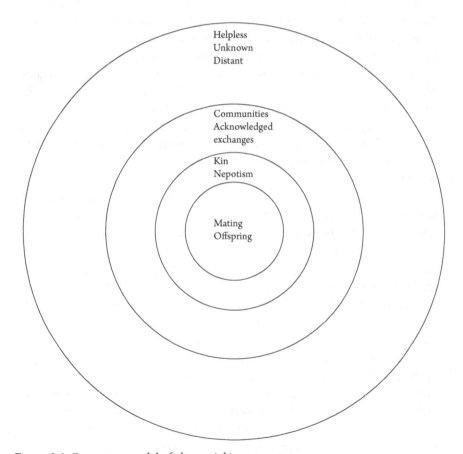

Figure 8.1 Four-zone model of ultrasocial investment.

As described in Chapter 7, efforts and actions within the heavy dividing line have direct genetic consequences on the biological variable, fitness, whereas those in the outer zones have consequences too, but indirectly and later, through others' actions in the context of one's reputation, or perceived "place" in the community.

Here, Alexander calls attention to the problem that arises: put simply, that no one can put effort into all these things at once, partly due to limitations in time, space, and available effort, but especially because they can come into conflict. Granted, each zone itself is full of difficult decisions, not least issues surrounding mate choice, but the bigger point is conflict between whole zones: when energy invested in one of them requires sacrificing a potential investment in one of the others. I like to call this shifting, from any zone to any other:

- Shifting-in: decreasing or ceasing effort in any zone except the center, to invest more in one or more zones inside it
  - The community zone's complexity can be confusing, but to shift inward from it is identifiable because relatedness is the boundary. It's no surprise that many communities co-opt the language of "family" to blur it, claiming either that the community is itself a family or that it represents the interest of families.
- Shifting-out: decreasing or ceasing effort in any zone except the outermost, to invest more in one or more zones outside it
- Shifting all the way out, such that all of one's investment is identifiable in the outermost zone, is an extreme experience for the acting person and for observers.

Any given instance requires more specification, such as which community is being favored or denied, or which kin member, or any number of other subtleties, but the general point stands, that such decisions are loaded with a distinctive form of stress and a distinctive set of social risks—as if we needed an additional source of stress and risks on top of those listed in Chapter 7.

This model entirely sets aside the usual focus on motivations and does not identify "right and proper" action. Any choice within a zone or any shift among zones may be assessed as moral *or* immoral acts, depending on the action and how it is received. The point is to explain why we experience certain conflicts and stress in a particular way that we refer to as "moral." The stress, especially, arises from simultaneous prompts for incompatible actions; basically, whatever you do has some element of price or sacrifice, and quite likely, someone is going to call it wrong and do something about that.

The breakdown of religion that I provided is practically a road map for negotiating one's way through these problems. Observance establishes the primary connection between one's kin and a designated community, and that designated community is positioned in the larger picture of economics and historical power.

Therefore being "born into" a particular religious practice establishes a social position and a set of alliances, as well as specific risks.

Metaphysical belief doesn't have much to do with it, which is tough to talk about because "belief" is also used to describe commitment to the community. The word seems almost designed to confound reflection about the difference between religiosity and the other components, such that even proclaiming the belief itself, as in "I believe" or "God is great," becomes a tag for specific social positioning such as militant resistance, military action, and institutional oppression and genocide. I do not think that the religiosity component of such statements uniquely *makes* anyone do anything, absent those social and political conditions that lead to such actions anyway.[3]

There's always a tension among these components, as institution and doctrine represent powerful third-party alliances at the community level, reaching into the center of the diagram to affect parenting. Also, if the influential members of a community or institution can align the language of these different components, they can tap into the forces of astonishing collective action, especially when the community is undergoing internal or external stress. The novel captures this point in remarkably few words when the Beast Folk are shocked into action by the Leopard Man's attack upon Moreau—but toward what, is briefly up in the air. Until Moreau's pistol shot redirects them, it looks as if they're about to focus upon Prendick as the nearest source of confusion and instability.

Alexander also suggests that institutional religion often features a focused exploitative tactic: older individuals influence the learning of the young, such that the younger individuals tend to shift-out more often than they might otherwise do. The training often dramatizes the most extreme shift-out with the portrait of an individual who sacrificed direct investment (the innermost circles) in order to benefit the community or humanity at large. It also features regular rewards for shifting-out toward the teaching institution. Therefore doctrinal education includes a social tactic to perpetuate the power structure in which the elders' fitness-potential is embedded, minimizing the chance that younger individuals will overturn it. It is effectively disinformational education, such that by the time people shake off the training, they will be too invested in the existing societal structures to pose a threat to them.

Most of Huxley's clashes with Anglican scholars and politicians arose because they were defending their institutional privilege and control. Those privileged in the institution claim metaphysical justification, whether you call that God, the cosmos, or nature, by co-opting the extant religious practices and phrases. Messing with such justifications is, to them, an attack on their power—which is also a moral breach, as that is what they have been teaching all along. Their indignation may well be sincere, but even as they insist that the discussion is about morality and social order, it is very much about punishing dissenters with character assassination, force, misery, and pain.

When Moreau calls the Beast Folk together and invokes the Law in the context of a criminal court, he shows that he knows about the Law a bit more than he had allowed to Prendick during their debate the previous evening. He and Montgomery have institutionalized the Beast Folks' observance into an authoritative structure they want to preserve, adopting their community rituals into church doctrine and effectively into civic law, backed with their monopoly on condemnation, force, and pain. This was the source of their anxiety in Chapter 13, when Prendick—thinking the Beast Folk are altered humans—calls for them to organize against their oppression. If the Beast Folk were utterly programmed as Montgomery claims, or were really "only" beasts as Moreau dismisses them, then this would make no sense at all, but as usual, the two understand the Beast Folks' human and cultural features strictly pragmatically, without acknowledging them further.

Hell Is Real

When the Beast Folk talk of the House of Pain, it means both the laboratory as a physical building and a cosmic concept. There is the lab, but there is also its meaning for them, an essential component of their view of the universe, which they struggle to apply to their lives and include in their rituals. Moreau has taken this over by using the lab as a social threat, an enforcement mechanism for their observance and for their collective submission.

Whether Moreau understands this is not clear, as he seems cynical about it, and it's possible his description of the Leopard Man as the "sinner" is sarcastic, for instance. However, as I described earlier, he certainly applies the same strategy against the Beast Folk as religious institutions apply against real people. It wouldn't be the only way that Moreau reacted to and manipulated the Beast Folk as people without giving them credit for that status. My interest lies less in him than in them.

In practice, people don't experience belief, observance, and institutions as independent. Experientially, one internalizes what was one taught, such that personal belief about metaphysical consequences, day-to-day practices, and institutional enforcement are all intertwined regarding how to behave. Although Alexander focuses on the nicer content of what children are taught, the standout example in the history of such things is certainly the threat of punishment. In the Abrahamic tradition most relevant to the novel, we're talking about Hell. I don't mean Hell as a metaphor or as a trivial descriptive device, but as a deeply felt and relevant place of agony that one fears, as a possible consequence for having misused one's life. Early indoctrination and generalized social compliance founded on this idea are serious business, understandably affecting people's behavior because, although they can never see this place, they certainly know what pain and misery are.

That's why Hell is such a vivid and significant component of Abrahamic doctrine, even though it's based on the thinnest possible textual foundations, dubious extrapolation from equally dubiously inserted text. It's built up in secondary texts (Talmud, catechism, hadith), especially those directed toward education, adding imagery, details, characters, and mythological events in ever-escalating profusion. No Abrahamic religious institution wielding government-level social power ever dialed down its imagery of Hell, and as Alice Turner describes in *The History of Hell* (1995), even the most rigorous revisions of doctrine in institutional schisms have not only preserved but also reinforced the role of damnation, completely counter to the schismatics' purported goal of "getting the text right." This effect has been repeated through multiple revisions and reformations, suggesting that eternal torture's social utility outweighs attention to metaphysical and textual consistency:

- Orthodox Judaism (mystic): *Gehinnom* (waiting, punishment)
- Orthodox Catholic Communion (Constantinian/Augustinian): *Gehenna* (Day of Judgment, eternal torture)
- Islam: *Jahannam* (immediate eternal torture)
- Roman Catholicism: *Infernus* (Day of Judgment, eternal torture, penance)
  - Lutheranism: *Hölle* (immediate eternal torture)

In the nineteenth century, these elements underwent some changes in both establishment and marginal forms of English-speaking Protestant Christianity, the bigger context of profound sociopolitical change, in which the roots of today's policy battle lines are sometimes visible and sometimes thrown into confusion. Among the sea of dissenting and organizing movements, the term "Chartist" applied to a wide range of causes, often influenced by French Revolutionary events, often invoking evolution specifically in terms of Lamarck, and especially dedicated to broadening the base of representation in Parliament (in Figure 2.2 in Chapter 2, Chartism would be Evolutionary, Revolutionary, and Reform—a potent combination). The derogatory word "establishment," with its connotations of cronyism, privilege, and authoritarianism, dates from this period.

The establishment, then, was well situated to resist and suppress the Chartists and related threats, not least in the power of the Church of England. But the Church was changing, too. As its reach expanded geographically through the spread of empire, it fragmented into new dioceses with their own local priorities. Domestically it was gradually reshuffled into High, Broad, and Low practices, respectively summarized as authoritarian and rather medievalist, intellectual and non-literalist, and evangelical or pietist. United States religion underwent historic changes as well, as the Episcopal Church would become the most politically powerful establishment church there, specifically the "P" in the term "WASP."[4]

As for Hell, church intellectual culture was marked by intensive and bewildering debate throughout the century, quite a bit of it opposed to the concept of a literal place where bad souls writhed in agony for eternity. In practice, however, the imagery remained in place, notably in education. Roman Catholicism especially may be said to have doubled-down on the concept, with a Papal Bull in 1879 to that effect, and the established Protestant churches as well. Irrespective of what the rarefied scholars were saying, a newly emphasized and ever more lavishly illustrated damnation was made clear to observant churchgoing people.

So much for the big churches and governments. Something else was afoot, too, in the drama of radical Protestantism, undergoing organizational paroxysms and cultural redefinitions that would come to define modern social representation and political activism. Its roots lie back in the 1730s and 1740s, mainly in the southern United States, sometimes called the First Great Awakening: a focus on ecstatic practices and local doctrinal interpretations within local branches of established churches, as well as the origins of the Baptist and Methodist Churches. From the 1790s to the 1840s, this tradition further developed the new post-millennial Restoration Movement, leading to a unique doctrinal culture in the Baptist, Methodist, and Presbyterian churches, and the origins of many new sects. This movement was associated with the Second Party System in the early to mid-1800s and represented a powerful new political development across rural North America, and is referred to by modern evangelical culture as the Second Great Awakening. It overlaps with a groundswell in social activism, abolition first among its parts, also temperance and a great deal of electoral and land reform. In the second half of the nineteenth century, the sometimes-called Third Great Awakening combined these churches and trends with continued social activism, economic reform, and mysticism. This movement—better described as a number of disconnected groups—often identified with "science" at least as a term, in an uneasy relationship among political organizing, mystical intensity, and modernism, including the relatively socialist branch, the Social Gospel, and the spiritual-healing branch, New Thought.

A fair number of these socio-religious organizing movements included talk of evolution in the form of the generally teleological Christian Darwinists: Joseph Cook's *Biology, with Preludes to Recent Events* in 1887, the popular American preacher Henry Ward Beecher's *Evolution and Religion* in 1885 with its "sublime march toward perfectness," and Lyman Abbott's *Evolution of Christianity* in 1892, seeking a religion "cleansed of pagan thought." Events in the United States were soon to evict all talk of evolution from evangelical Christianity there, but until around 1900 about half the American churches in this category were explicitly pro-evolution in some form.

And what did Hell have to do with it? A great deal in the evangelist rhetoric, with a fervent and enthusiastic invocation of fiery torture straight out of the Middle Ages Catholic and American Puritan playbooks. The most progressive, least overtly biblical version is quite subtle. Cobbe's thoughtful *Darwinism and*

Morals (1872) is by no means a stereotyped fundamentalist tract. She specifically condemns Darwin's ideas about the material origin of human values, citing only Repentance and Regret (her capitals) at a spiritual, cosmic level to be the only means by which virtue is made possible. It's stated in measured tones without the special effects, but the implied punishment—for which she uses imagery from ancient Greek drama—has much in common with the shame and dread invoked by devils in *Gehinnom*, and her whole point is that it is not based on material psychology and social dynamics, but on genuine metaphysical content.[5]

In the novel, Hell is never mentioned by name. Moreau comes closest to discussing it explicitly, in his debate with Prendick in Chapter 14, when he rudely but accurately equates the vague Christian heaven with his orientalist phrasing, "Mahomet's houri in the dark." He is pointing out the crudity of Abrahamic religious morality: virtuous behavior is rewarded with ecstasy and the implicit withholding of agony. Clear away the special effects and all you see is a rather crude psy-op, nothing "higher" about it. His view is related to his contempt for the Beast Folk's "not Man enough" status, in that they can be bullied by such methods, and to the definition of materialism he despises.

The infernal presence is certainly felt, though, in multiple scenes, for multiple characters, and in multiple conflicts. Details of the Beast Folk's plight are distributed across characters and events, reducible to the chilling fact that, unlike people everywhere else, the Beast Folk *know* that Hell is real. It's a podium-thumping evangelist's and a social engineer's dream come true: if only those wretched misbehaving sinners would feel the genuine touch of hellfire right here and now! *That'd* teach them.

In a 1979 academic paper, *The Island of Doctor Moreau as Theological Grotesque*, Gorman Beauchamp nailed it: although Moreau isn't playing God, the God of the nineteenth-century European church establishment is apparently a whole lot like Moreau. His insight regarding punishment snaps together well with Alexander's model of religious indoctrination as a specific form of exploitation, tweaking people's existing capacity for religiosity.

The whip is a powerful visual and social object in the novel, although it is used more for attention and direction than for inflicting pain directly. If the few instances of its use or reference are anything to go by, the Beast Folk don't even particularly fear it. The films have utilized it more toward this latter purpose, especially in *Island of Lost Souls* (1932), *The Island of Doctor Moreau* (1977), and—translated into a remote-activated electronic implant—*The Island of Doctor Moreau* (1996). Each of these films places Moreau as the Law's author, which changes the meaning drastically. Instead of Moreau's whip co-opting the Beast Folks' observance of their grassroots religion, the pain-inflicting weapon is itself the founding and defining component of the religion they're forced to practice.

Huxley's *Evolution and Ethics* ties into this point, although ambiguously—in some parts he describes the mismatch between common human values and civil society, including the latter's equally brutal enforcements ("rope and axe") against them; in others, about recent improvements in human ethics. Nothing in his portrait supports a notion of existing human ethics in widespread, applied and socially functioning form, and his "rope and axe," to which one can add a long list of implements and coercive threats, is still the primary method of attempting to curb ordinary human behaviors.

The Leopard Man's plight is a letter-perfect dramatic version of Huxley's verbal struggle with these concepts. He "goes under," not in the sense of reverting to inhuman bestiality, but rather in terms of simply giving in to existing urges, period, and his institutional religion cannot help him, only punish with its crucial blend of pain and shame. He is an observant who has insulted his community, and he's a believer, too, so he cannot genuinely rebel or speak against the institution.

> Moreau stopped, facing this creature [*the Leopard Man*], who cringed towards him with the memory and dread of infinite torment. (*The Island of Doctor Moreau*, p. 70)

Look carefully at his attack on Moreau—here we have one of the two most physically dangerous Beast Men on the island, certainly capable of killing a person as easily as you or I might open a banana and with much the same motions, and all he does to his surprised target is knock him down, before fleeing—with nowhere to go. It's not rage or bestiality, but extreme fear that leads him to rebel, and within his rebellion lies abject despair. The old spiritual from the 1920s could well have been written for him:

> There's no hiding place down here / There's no hiding place down here
> Well, I run to the rocks to hide my face / The rocks cried out, no hiding place.

When Prendick finally sees it, he instantly emphasizes; the gap between them is narrowed and vanishes. Enlightened freethinker he may be, but he's "no materialist" either, and here he demonstrates in no uncertain terms that he knows all about the *bad behavior = damnation = agony* equation.

The Jesus Moment

Sometimes, it seemed to me as if my literature teachers shared a veritable fetish about what they called "the Christ figure." *Lord of the Flies*—Simon's the Christ figure! *One Flew over the Cuckoo's Nest*—McMurphy's the Christ figure! And so on and on, until I came to suspect that not even Winnie-the-Pooh or Conan the Barbarian might be safe. My frustration with the concept arose in part due to its apparent range in application, but mainly because no one ever discussed what this "figure" might be *for*, or what it contributed to the story as such, beyond providing a highbrow *Where's Waldo* for these classes.

Therefore I present the following idea only reluctantly, and with dread that some one of those bygone instructors, now quite elderly, is about to appear on my doorstep and say, "See! I told you!" But here goes: Prendick becomes Jesus. Or more accurately, the Beast Folk find Jesus.

Let's take a look at the interesting views toward this personage or icon in the late 1800s. The Roman Catholic Church and the branches of the Anglican Communion, including the Episcopal Church in the United States, were finding new roles in modern forms of political power, with the former two somewhat diminished and the latter becoming quite significant, but were all still essentially medieval in outlook toward Jesus: a vague, perfect, not especially volitional image who was most relevant either when in agony or when ascended out of reach. The older continental Protestant churches like Lutheranism and their US derivatives were similar; their initial insistence that salvation was a personal matter had worn thin over the centuries, and they'd never focused much on Jesus' sayings and doings anyway. As for the newer raft of doctrinally radical Protestant sects, they had a lot to say about loving Jesus and being forgiven by Jesus, but similarly, not so much about what one might think of as plot or character, except perhaps among the Populist movement in the United States, where "moneychangers" received a good verbal roasting in the Chautauqua tents.

However, something new had arisen among middle-class Christians, especially those engaged in progressive activism: the notion that if you were to scrub away all this "church" talk and accumulated institutional doctrine, you'll find less Christ and more Jesus, who happens to be a socially aware, anti-authoritarian, countercultural person. The New Testament received a rehabilitation in this context, becoming a dramatic narrative starring a principled, kind person with understandable emotions, and whose suffering connoted injustice rather than a cosmic guilt-trip. The idea was also implicated in a new kind of biblical scholarship, which dates from this period in the work of David Strauss: the dissection of religious texts in order to distill them to their essential or original forms (eventually this endeavor would lead to the discovery that there were no such original forms, but that is another issue).

The renewal of Unitarianism during the 1820s in the United Kingdom and the United States and its transition into abolition and other political efforts are the perfect landmark for this idea, but it was present as well in branches of existentialist philosophy, Transcendentalism, and socialism. It differed from evangelical tradition in its complete dismissal of an afterlife and Hell, but was similar to it in some political terms, with one foot in radical ideas and one in legislative reform.

Evolutionary terms were a big part of this development, too. Both the novel's national origin and its timing reflect how intellectual Christianity, in its activist manifestation, favored evolutionary talk in a variety of combinations, just as evangelical Christianity did. It was often called "Darwinism," but drew less from Darwin than from the general and improvement-oriented model from the *Vestiges*, from Huxley's optimistic views of science as the means of human liberation, and from readings of Spencer. Good examples include the Scottish author and prominent figure of the Free Church of Scotland, Henry Drummond, in his books *Natural Law in the Spiritual World* (1883) and *Ascent of Man* (1894). The Unitarian minister Minot Judson Savage produced numerous works on his concept of "Darwinian evolutionistic optimism" for three decades, with titles like *Christianity, the Science of Manhood* (1873), *The Religion of Evolution* (1876), *The Morals of Evolution* (1880), and *The Evolution of Christianity* (1892).

This image of the dissenting, organizing, gentle but firm Jesus was very effective in progressive and legislatively aimed activism among scholarly and middle-class circles, in which long doctrinal discussions tend to play a big role. Religion is portrayed as getting better through Darwinist processes, casting off bad or vestigial features, literally evolving a new and better social application, in part by discovering and seeing the "real" Jesus. Conveniently, one could self-identify as either religious or atheist, because the sense of historical-theological "Jesus-ness" can sit nicely in your social activism whether he's the son of God or not, or if there is or isn't a God.

Consider as well the contemporary European views of Judaism, which were generally negative, especially toward obvious observants. At worst, Jews were historically vilified as the Christ-killers and at the least viewed as cultists almost certainly up to some scheme or another. This outlook could be invoked as politically expedient, and it often was. Doctrinally, Christian groups and speakers universally considered Judaism as providing nothing more than raw material for the real religion, represented by Jesus, himself the first Christian—there was no talk of "Judeo-Christianity" then. They tagged Jews as the "blind who will not see," stubborn and obscure pagans who unaccountably clung to their manifestly primitive precursor rites to their awful obsolete tribal deity, refusing to get with the clearly decent and reasonable program, itself uncritically perceived as having done so much to make humanity better.

This viewpoint is evident among all the established Christian churches, but it was held across the progressive movement and emergent sects as well. Savage provides a remarkable summary in *The Morals of Evolution* (1880), which incidentally includes a dedication to Spencer, with his condemnation of both Judaism and institutional Christianity:

The Hebrew god at the first was never conceived of as a moral being at all. . . . Think for a moment of some of the characteristics he manifested.

He taught Jacob to steal by fraud and falsehood the birthright from his elder brother. . . . What kind of a god was it that led Joshua to the conquest of Canaan, directing that he slaughter without mercy man, woman, and child of city after city; that they rob and take possession of the whole country? [*five similar examples follow*] Are these moral conceptions? Is this a moral religion of which such things as these are a consistent part?

. . . Is the religion of Jerusalem at the time of Jesus a moral religion? Does not Jesus himself say that they have made the temple a den of thieves? . . . that this [*hypocrisy*] was their religion and their practical life was a falsehood, was a deception, was the devouring of widows' houses? This was the kind of a picture that Jesus draws for us of the Hebrew religion at its culmination at the time when Christianity was born. (Judson Minot Savage, *The Morals of Evolution*, pp. 27–29)[6]

Christian institutional history gets it in the neck too, without even a paragraph break:

Who was the ideal and model Christian in the early ages of the Church, and through the Middle Ages until modern times? Who was he? Was he a man who was honest, a man who was right in his relations with his fellow-men, a man who did all that he could to build up human society, a man who illustrated the essential, fundamental principles of morality? Not at all. He was a man that lashed himself, that repeated so many aves, a man who said so many prayers, a man who fasted so many times a week, a man who devoted himself to the ritual and ascetic side of life, being of no use whatever to the world, and only being religious that he thus might gain power with heaven and secure his own selfish felicity in another world. (Ibid., p. 29)
[*he continues with a similar indictment of modern churches*]

I stress this presentation as characteristic of the full religious-political spectrum. In the case of the humanist or progressive versions, those Jews' weird and bloody tribal-god was over, and so was the remote yet sadistic oppressor-god of the medieval church. Now that they've found the right Jesus, a person who can be liked and respected, religion can be reformed into an engine for social progress and society can evolve into a new and better form.

The Beast Folks' religious experiences display this perceived societal drama in detail. They observed Prendick keenly throughout his first days on the island. They heard him call to them to defy Moreau in no uncertain terms. They saw him bleed and weep, leading them to debate his identity relative to themselves:

"Was he [*Prendick*] not made?" said the Ape Man. "He said—he said he was made."

[*and*]

"Yesterday he bled and wept," said the Satyr. "You [*Montgomery*] never bleed nor weep. The Master does not bleed nor weep." (*The Island of Doctor Moreau*, p. 65)

Montgomery's threats fail to resolve or to silence their discussion, which covertly continues among the Beast Folk throughout this part of the story.

Then they see Prendick defy Moreau again and save the Leopard Man from the House of Pain. If they discuss this—and it's hard to credit they don't, with the Ape Man and the Satyr around—Prendick doesn't hear it or report it.

Six weeks later, when Moreau is killed, the Beast Folk are acutely distressed, as initially dramatized to a Nietzchean degree over his body:

"He is dead," said a deep and vibrating voice. (Ibid., p. 79) [*soon identified as the Sayer of the Law*]

The ensuing discussion is fully theological—what does this mean?—and ethical—now what do we do? The Ape Man doesn't produce his customary chatter. He and the others get right to the point.

"Is there a Law now?" asked the Ape Man. "Is it still to be this and that?"
"Is there a Law?" repeated the man in white. "Is there a Law now, thou Other with the whip? He is dead," said the hairy Grey Thing. (Ibid., p. 80)

Prendick's own narration states that there is "nothing threatening" about the stricken group. They care deeply about this matter and are asking for help.

However, since he is consumed by fear that the Beast Folk might perceive the event as a cue to attack him and Montgomery, he offers an interpretation that he hopes will keep them at bay. He says, no, Moreau's not dead, he's invisible and watching, and now I am here instead. The Sayer of the Law gradually accepts Prendick's direction, culminating in shifting his address of him from "the Other with the whip" to "Man who walked in the Sea."

Prendick's tactic is quite cynical, seeking to keep them Law-abiding, or as he sees it, less inclined to attack him and Montgomery. But never mind Prendick's point of view for the moment. Look at where the Beast Folk are coming from when they buy it, what content is carried by accepting Prendick's concepts and interpreting his physical presence among them. Soon after these deliberations, he releases all the subjects in the House of Pain from their torment and then dramatically destroys the whole thing, and although he did the latter by accident, they don't know that.

The depth of their understanding isn't trivial, either. In the further dialogue after the deaths of Montgomery and the Sayer, I don't find much to criticize:

> "He is dead, he is dead! the Master is dead!" said the voice of the Ape-man to the right of me. "The House of Pain—there is no House of Pain!"
>
> "He is not dead," said I, in a loud voice. "Even now he watches us!"
>
> This startled them. Twenty pairs of eyes regarded me.
>
> "The House of Pain is gone," said I. "It will come again. The Master you cannot see; yet even now he listens among you."
>
> "True, true!" said the Dog-man.
>
> They were staggered at my assurance. An animal may be ferocious and cunning enough, but it takes a real man to tell a lie.
>
> "The Man with the Bandaged Arm speaks a strange thing," said one of the Beast Folk.
>
> "I tell you it is so," I said. "The Master and the House of Pain will come again. Woe be to him who breaks the Law!"
>
> They looked curiously at one another. With an affectation of indifference I began to chop idly at the ground in front of me with my hatchet. They looked, I noticed, at the deep cuts I made in the turf.
>
> Then the Satyr raised a doubt. I answered him. Then one of the dappled things objected, and an animated discussion sprang up round the fire. Every moment I began to feel more convinced of my present security. I talked now without the catching in my breath, due to the intensity of my excitement, that had troubled me at first. In the course of about an hour I had really convinced several of the Beast Folk of the truth of my assertions, and talked most of the others into a dubious state. (Ibid., p. 94)

Prendick sneers at their credulity, but he reports nothing actually stupid or which indicates dumb-brute acceptance. They ask questions, they talk about it, and they work it out. Furthermore, look at what the Dog Man reports about their conclusions, which is not what Prendick had been trying to achieve at all:

> We love the Law and will keep it, but there is no pain, no Master, no Whips ever again. (Ibid., p. 93)

Here are both of the transitions as conceived in the late nineteenth century: shifting from a demanding, bloodthirsty, arbitrary, and punishing God, to a more personal, more sympathetic, more merciful, reconciliatory, intercessionary and ambiguously human/divine version. They even thought of him as one of them at first, then accepted him as a higher—but different and kinder—divine being. But the next day, he demands more—to become a new institution of his own—effectively replacing the Pharisees with the medieval Church, shifting the whip from one name to the next, and they are saying, "No."[7]

The Beast Folk continue to practice the Law during the ensuing nine months, prior to their reversion. Unfortunately, Prendick reports little about how they do so, whether unchanged or altered by their reinterpretation of his presence and his claim regarding Moreau still watching them, but from what he does report, including that they behave "with decorum," they have perfectly displayed the late nineteenth-century progressive understanding of historical Christian development, as conceived by writers like Savage.

My thinking is that in this case, the Christ figure is indeed present in terms of *plot*, that is, what characters do, what characters observe, and how they choose to act upon it, as opposed to merely being "there" for some obscure reason. Not only does it repeat the contemporary narrative of the transition from one form of religion to another, but also matches with the relatively new, politically progressive notion of Jesus as an inadvertent historical figure, and that the circumstances of whoever Jesus was or whatever he did is eclipsed by the story that has arisen and become instructive. It nails down the Beast People's identity as human people at the level that is often tagged as a primary dividing line for human exceptionalism.

It's also Huxley's *Evolution and Ethics* insofar as it focuses on religion as a cultural ritual, how people feel and what they do, without special goodness attached to it (unlike Savage, who seeks for religion indeed to result in goodness). It's not expected to be factually correct, insightful, or even sensible, because it's at best a coping mechanism, partially effective only because life itself is so bad. The historical details of the new creed's origin are even rather brutal, insofar as Prendick's method of divine mercy is euthanasia by bullet. The portrait is exquisitely precise, acknowledging the presence and possible necessity of human religion in a world of unceasing suffering, but also striking directly against the dichotomy between uplifting, enlightening, spiritual human qualities that counteract terrible, suffering animal traits.

The Rebel

It's probably easy to see the Hyena-Swine much as Prendick does: a complete bastard, criminally murderous when he can avoid punishment, simpering and pious when observed, a skulking, malevolent hypocrite. I suggest reconsidering this judgment in light of their direct opposition following Prendick's second round of metaphysical claims upon Montgomery's death, which is much more directed

toward establishing a new institution and includes his demand for personal obedience. He deserves special mention as the only Beast Man who defies the Law
and its self-appointed authorities strictly on principle:

> . . . I heard a light footfall behind me, and turning quickly saw the big
> Hyena-Swine perhaps a dozen yards away. His head was bent down, his
> bright eyes were fixed upon me, his stumpy hands clenched and held
> close by his side. He stopped in this crouching attitude when I turned,
> his eyes a little averted.
>
> For a moment we stood eye to eye. I dropped the whip and snatched
> at the pistol in my pocket. For I meant to kill this brute—the most for-
> midable of any left now upon the island—at the first excuse. It may seem
> treacherous, but so I was resolved. I was far more afraid of him than of
> any other two of the Beast Folk. His continued life was, I knew, a threat
> against mine.
>
> I was perhaps a dozen seconds collecting myself. Then I cried:
> "Salute! Bow down!"
>
> His teeth flashed upon me in a snarl. "Who are *you*, that I should—"
> (*The Island of Doctor Moreau*, p. 89)

These are the only words from the Hyena-Swine reported in Prendick's entire
account. His dialogue is cut short not due to inarticulacy, but by Prendick's firing at him. Brief as it is, it cannot be characterized as imitative or rudimentary.
Grammatically, he uses the subjunctive better than most English speakers. Nor is
it some bestial grunt of unthinking defiance: Who can blame him for not wanting
to grovel, or for questioning anyone's claim to godhood?

Sadly, this curtailed dialogue is all we've got. But it joins the Leopard Man's
"No!" and the Puma Woman's virago-scream as a moment in the story for a voice
that neither Moreau as God-figure nor Prendick as his presumed heir can control.
The Hyena-Swine takes it to a new level, as his situation is neither encapsulated
and resolved as with the Leopard Man, nor restrained in the background as with
most of the Puma Woman's story. Like her upon her escape, he's right there, neither cringing nor captive, making his point. Unlike her, he wants an answer. What
led the Hyena-Swine to this moment—what has *his* story been up to this point in
the novel?

- He's briefly introduced in Chapter 15 as one of the two most physically formidable Beast Folk, which after the Leopard Man's death leaves him at the top of
  that hierarchy.
- Prendick accurately spots him to be the Leopard Man's rabbit-killing crony in
  Chapter 16. Throughout the following events, he lets the Leopard Man take
  the rap and, unlike him, apparently enjoys defying the Law. He hides it with

ostentatious observance at the convocation called by Moreau and during the disposal at sea of the Leopard Man's body.

- He is not present during the Puma Woman's escape and Moreau's death, nor during the drunken brawl and Montgomery's death.
- In Chapter 20, he defies Prendick's claim to divine authority and is shot in the hand by him.

I think it likely that he was aware of the enforced Law's hollowness, relative to Moreau, since before his introduction into the narrative. When Prendick seeks to re-establish Moreau's power, albeit more subtly, the Hyena-Swine defies it in correspondingly more subtle form, as the only Beast Man in the story to consider his or her circumstances in late nineteenth-century philosophical terms. He becomes the dedicated modern, openly disgusted by doctrinal authority and its claim to divine justification.

Earlier in the story, Prendick gave voice for what life is like for the Beast Folk, restricted to some parts of Chapter 15 and to the end of Chapter 16, at the funeral of the Leopard Man. In this he provides, without knowing it, the context for this later confrontation—what he said then is exactly what the Hyena-Swine is now acting upon. Specifically, at the end of Chapter 16, almost word-for-word from Huxley's description of humanity in *Evolution and Ethics*:

> Before they had been beasts, their instincts fitly adapted to their surroundings, and happy as living things may be. Now they stumbled in the shackles of humanity, lived in a fear that never died, fretted by a law they could not understand; their mock-human existence began in an agony, was one long internal struggle, one long dread of Moreau—and for what? It was the wantonness which stirred me.
>
> . . .
>
> I must confess I lost faith in the sanity of this world when I saw it suffering the painful disorder of this island. A blind fate, a vast pitiless mechanism, seemed to cut and shape the fabric of existence, and I, Moreau by his passion for research, Montgomery by his passion for drink, the Beast People, with their instincts and mental restrictions, were torn and crushed, ruthlessly, inevitably, amid the infinite complexity of its incessant wheels. (*The Island of Doctor Moreau*, p. 74)

This is familiar territory to students of philosophy. It is the none other than the Absurd, straight from contemporary pre-existentialism, and although I am no scholar of every last text, to me it evokes this bit from Søren Kierkegaard's *Repetition* (1843):[8]

> How did I get into the world? Why was I not asked about it and why was I not informed of the rules and regulations but just thrust into the ranks

as if I had been bought by a peddling shanghaier of human beings? How did I get involved in this big enterprise called actuality? Why should I be involved? Isn't it a matter of choice? And if I am compelled to be involved, where is the manager—I have something to say about this. Is there no manager? To whom shall I make my complaint? (p. 200)

In regard to religion specifically, the concept also corresponds exactly to the later essays of Sigmund Freud, specifically *The Future of an Illusion*[9] (1927) and *Civilization and Its Discontents* (1929). It's not quite fair to quote works that post-date the novel here, but since Freud's younger self does show up in its final chapter, I'll cheat:

> A believer is bound to religion by certain ties of affection. But there are undoubtedly countless other people who are not in the same sense believers. They obey the precepts of civilization because they let themselves be intimidated by the threats of religion, and they are afraid of religion so long as they have to consider it as part of the reality which hems them in. They are the people who break away as soon as they are allowed to give up their belief in the reality-value of religion. (*The Future of an Illusion*, p. 59)

There's a word for someone in this situation: alienated. The Hyena-Swine nails the concept to the wall in his single reported phrase. He sees the dishonesty of the moral institution he lives in, hates it, and mocks it. He does not fear Hell, enjoys defying authority when it's too arrogant to see that he doesn't buy it, cannot be ferreted out by guilt or community pressure, and immediately picks up on the first evident crack in Moreau's and Montgomery's presumed godhood. When institutional power finally brings its force and pain directly to his door, he denies it its greatest weapon, internalized compliance.

Some touches of the Hyena-Swine show up here and there in the films. In *Island of Lost Souls* (1932), Ouran convinces the Sayer of the Law that Moreau is a diabolical God, and a similar sequence occurs in *The Island of Doctor Moreau* (1977), in which M'Ling aids the Sayer of the Law in confirming that Moreau has killed Montgomery. The latter film also features a slight religious disagreement when the Bull Man defies the Sayer's plea to suppress aggression.

The literal hyena character in *The Island of Doctor Moreau* (1996) is called "Hyena" in the dialogue, although credited as Hyena-Swine, played with energy and sympathy by Daniel Rigney. If he and Douglas had more content scripted into their confrontation near the end, and if that sequence had

not veered into everyone-shoots-everyone, the overall story may have gelled
more successfully. The film contained elements of his deliberate reversal of
the Law and some paramilitary imagery, which could have factored into that
idea pretty well.

In *Dr. Moreau's House of Pain* (2004), the corresponding character is called
"Peewee" (B. J. Smith), composed of hyena and mountain lion (i.e., puma).
He is played mainly as a brute, killing women and carting around their bodies like dolls, although with a surprisingly poignant end after his friend is
killed.

Prendick's access to the voice for the Beast Folk is over very quickly. At the
Leopard Man's funeral, his state of mind was "morbid," but also "alien to fear."
Now, six weeks later, that state of mind has vanished, and in this moment, his
insistent fear that the Hyena-Swine is "about to revert" is all he thinks about. Look
at him with his gun, his whip, and two hatchets, facing the Hyena-Swine on the
beach. The latter does not attack. He waits to see what Prendick will do and say.
When ordered to bow down, he still does not attack, he speaks with arguably righteous anger—and for this, he is shot and maimed.

The whole story flips over in this moment. Who is its protagonist, again? Not
Prendick at all. Prendick failed to listen to the Leopard Man, perhaps understandably, he failed to seek justice for the Puma Woman, which I find harder to forgive,
he failed to ask M'Ling for his opinion, which arguably cost him and Montgomery
their lives, and now he fails this guy, too.

For the next ten months, the Hyena-Swine goes to live by himself, as why would
he not, considering that an armed lunatic in the village is claiming to be God and
is willing to shoot him on sight, and who then escalates to trying to organize the
Beast Folk against him and failing that, to ambushing him with murderous intent.

The Leopard Man, the Puma Woman, M'Ling, and the Hyena-Swine: What
do these Beast Folk protagonists do? They are not heroic; rather, they are trapped
in an existence and a society for which metaphysical Hell is a pale symbol. They
writhe in agony both internally and externally, they rebel, they speak out and
are not heard, they remain silent because they know they won't be heard, they
try to put their thoughts together, they take refuge in action, they try to self-
isolate—and ultimately, they die, with the only question being whether it's on
their feet or on their knees.

Fiction has featured such characters for a long time, well before anyone made
up the word "existentialism." They're a broad bunch, hard to nail down into a
definition, ranging across acquiescence, isolation, humor, discovery of personal
ethics, savagery, and rebellion. Whether these responses qualify as heroic in any
given case is a matter for scholars; as a mere reader, I find them inspiring as well as
disturbing, even when the character is effectively a villain. Sometimes the whole

idea of heroism or protagonism can get vague in these stories, or is reduced solely to the character's unique recognition of the Absurd, in which any decision he or she makes loses traditional moral force but gains individual recognition.

Such characters range the whole spectrum from brave to cowardly, idealistic to cynical, and altruistic to vicious, and whatever moral judgment they invoke is typically not nicely unwrapped for the reader but left as a disturbing unopened package. I think the Hyena-Swine qualifies for inclusion in this tradition. Perhaps in the Green Room of literature, he, Medea, Edmund the Bastard from *King Lear*, Frankenstein's created man, Meursault from *The Stranger*, and Jerry from *The Zoo Story* are drinking together.

Readings

As with Chapter 7, my primary reference regarding human morality, religion, and evolutionary biology is Richard Alexander, *The Biology of Moral Systems* (1987), with some influence from Frans de Waal, *Primates and Philosophers* (2009). Most other writing about religion in this context is focused on religiosity, as with Boyer, *Religion Explained* (2002); Atran, *In Gods We Trust* (2002); and Trenton, *Minds and Gods* (2006); and one might hope for a reorientation toward institutions, policy, and community.

The interplay among politics, Anglicanism, and scientific ideas in which the Chartists, the Oxford movement, and dozens of other significant movements or cabals formed is described in detail by Adrian Desmond in *The Politics of Evolution* (1992). The Anglican priest Charles Kingsley bears special mention as possibly reconciling the institution with evolutionary thought almost single-handedly.

The scholarly reference for infernal regions is Geoffrey Rowell, *Hell and the Victorians* (2002), with an accessible version in Alice K. Turner, *The History of Hell* (1995). In reference to the novel, see Gorman Beauchamp, "The Island of Dr. Moreau as Theological Grotesque," *Papers on Language and Literature* 15(4): 408–417, 1979.

Francis Power Cobbe's *Darwinism and Morals* (1872) and related writings are a gold mine for discussion and reorienting modern assumptions about the topics.

The opening text for the investigation of events in the Gospels was David Strauss, *Das Leben Jesu* (1835–1836) (English: *The Life of Jesus*), with later impressive contributions from William Friedman, *Who Wrote the Bible?* (1987) regarding the Pentateuch (first books of the Hebrew Bible or Old Testament) and Burton Mack's *Who Wrote the New Testament?* (1996).

The interesting and generally unappreciated connections among evangelical Christianity and evolutionary thought are explained in David Livingston's *Darwin's Forgotten Defenders* (1984); James Moore's *The Post-Darwinian Controversies* (1981); and Bernard Lightman, "Christian Evolutionists in the United States,

1860–1900," *Journal of Cambridge Studies* 4(4): 14–22, 2009. The relatively abrupt transformations of this culture into a more centralized, one-message political movement in the United States is important but generally unappreciated history, involving the Granger movement and the subsequent Populist Party's fortunes in the 1890s. The on-message organizing, blanket doctrinaire consolidation, and the demonization and purging of evolutionary talk found their eventual textual unification in *The Fundamentals* in the early 1900s, which has led to the specific American meaning of "fundamentalist" and its explicit rejection of evolutionary theory. Useful texts about this topic include Daniel Williams, *God's Own Party* (2012); Darren Sochuk, *From Bible Belt to Sunbelt* (2012); and Karen Armstrong, *The Battle for God* (2001).

The philosophical school or trend discussed here isn't really existentialism, which didn't exist as a term until the 1940s, but earlier writers were often retroactively included, especially Søren Kierkegaard, *Fear and Trembling* (1843), *Sickness unto Death* (1849); Arthur Schopenhauer, *Parerga and Polimena*, available as *Essays and Aphorisms* (1851); and Friedrich Nietzsche, *Beyond Good and Evil* (1888). Twentieth-century existentialism isn't as relevant here, with the possible exception of Albert Camus's articulation of the Absurd in *The Myth of Sisyphus* (1942) and *The Plague* (1947). Literary theory regarding alienated protagonists is found in Colin Wilson, *The Age of Defeat* (1959) and *Introduction to the New Existentialism* (1966). They tend to focus on twentieth-century characters deliberately written into the philosophical discussion.

Notes

1. From 1869 through 1880, Huxley participated in the well-named Metaphysical Society, which was intended to bring scientific, religious, and philosophical views into civil dialogue. It was indeed civil, but perhaps for that very reason, yielded nothing much worth talking about except that no one knew what "metaphysical" meant.
2. "Kanaka" is a Hawaiian word adopted by British colonial culture and broadly employed toward Pacific and South Seas island workers in general. The original term may have "wild man" or "naked savage" implications, but as used colonially was typically derogatory, used much as "boy" in American English. A South Seas missionary from that culture would almost certainly have been Anglican, specifically Church of England, which is related to the new colonial dioceses established in the mid-nineteenth century.
3. Richard Dawkins's *The God Delusion* (2006) and related works from various authors are not relevant to this discussion, as they seem to me to confound religiosity with the other components of religion. They display a degree of category error I'm more accustomed to seeing in evangelical Christian political speeches, and their focus on religiosity, tagged with doctrinal phrasing, runs counter to all data-driven work concerning so-called religious violence. It also seems to me that Huxley's stark point in his original coining of agnosticism is largely missed: the issue is not whether there is a God, but whether one thinks that any such issue matters.
4. Much of the nineteenth-century Church story involved Ireland, including the joining and then splitting of the Churches of England and Ireland, and the closer integration

of the Roman Catholic Church with the High Church of England, and the first permission of Roman Catholics into British government since the Reformation, in the form of the Roman Catholic Relief Act in 1829. Politics and religion erupted in Ireland over the tithing of the whole population to fund the minority state church, leading to the Tithe Wars of 1834 and 1836, as well as the associated Irish Temporalities Bill in 1833. It was intended to relieve this pressure but was regarded with horror by conservative Anglicans, who favored a strong and unified state church across the British Isles. The Oxford movement represented this view, also called the Tractarians in reference to the pamphlets they published opposing the bill. The brutal and dismissive English response to the Great Irish Famine of 1845–1852 played into these developments, including the Irish Church Act in 1869, which permanently separated the Church of Ireland from the Church of England in 1871. Irish militant opposition to English political control began at about this time, developing through the Fenian movement in the 1880s and the revolt in the 1920s, leading to the partition of Ireland into two and then three separate nation-states as it stands today.

5. Cobbe's ethical position is evolutionary despite her rejection of Darwin's position, being most consistent with Wallace in exempting humans from the selective process. Unlike animals, the argument goes, humans have access to an external and objective meta-reality through an ineffable sense related to the metaphysical existence of beauty and religious sentiment. She disagrees with Darwin's idea that humans have a constant social sense as a physiological feature, citing instead religious repentance as a consistent brake on human rottenness, and also as evidence for her "sense of the sacred obligation of Rightfulness." Most relevant here, the overlap between or the common roots of evangelism and progressivism are explicit in her appeal to "what will stop the criminals unless they fear hellfire?" She grimly acknowledges that natural selection may be correct, but that acknowledging it would be socially dangerous, because people would act differently (i.e., badly) if they thought this way; in this, she references survival of the fittest as ruthlessness and lack of pity.

6. Judson Minot Savage, *The Morals of Evolution*, 1887, published by George H. Ellis, reprinted by Elibron Classics (Adamant Media Corporation), 2005.

7. The contemporary understanding of Abrahamic history was terrible, limited to an extremely diluted story of Jesus, then Roman Catholicism, then Protestantism, with no knowledge of the Orthodox Catholic Communion or the ancient diversity of Persian, Mesopotamian, Arabic, or African Christianity at all.

8. Søren Kierkegaard, *Repetition*, published by C. A. Reitzl's, 1843, translated by Walter Lowrie, 1941, published by New York Harper & Row, 1964.

9. Sigmund Freud, *The Future of an Illusion*, 1927, published by W. W. Norton and Company, 1961.

9

Beast Monsters

After Montgomery's death, Prendick lives for ten months with the Beast Folk, who display no signs of reversion, and then they revert all together in a few weeks' time.

This plot element might be a little baffling. Structurally, Chapter 21 seems like a 90-degree turn from the story as developed so far, which reached genuine climax at the end of Chapter 16 and ends with Prendick rather humbly joining the society of the Beast Folk at the end of Chapter 20. It even suffers from a circular quality, in that the Beast Folks revert because it's supposed to mean something, and their reversion is used as evidence for that meaning, which is a bit of a structural weakness, but also merits special attention when we're talking about themes.

It's easy to be distracted by Moreau's famous line about "the stubborn beast flesh" and his and Montgomery's constant fear that the violent urges of the Beast Folk are too strong to be repressed for long, or without forcible reinforcement. If their fear is accurate, then halfway through, the story looks like it will correspond to *Lord of the Flies* with different variables, because "the animal will out and there's nothing you can do about it," and also to correspond to the "it won't work because it's abominable" component of the "don't meddle" story.

In their fears, their concept of reversion is confounded among three extremely different things:

- Moreau's changing perception of each Beast Person following its transformation, which has nothing to do with the subject but with his overly idealized gaze.
- A given Beast Person's potential expression of violence, in stressful circumstances that are often accessible to a human's empathic understanding.
- Literal reversion into the animal from which a Beast Person was derived, including changes in posture, loss of speech, and disregard for the social norms of the Beast Folk.

These distinctions are confirmed, and the stated fears fully subverted, by key plot points. The human characters misinterpret the first two points through their imagined terrors and their blindness to the human qualities of Beast Folks' actions that they happen not to like. When the reversion does occur, it happens all at once and rather swiftly for all the Beast Folk. The reverted Beast Folk don't go on a rampage—but *Prendick* certainly does.

The Taste of Blood

What Montgomery and Moreau say is a lot like one might expect in the "don't meddle" story: "If they taste blood, they'll revert to true beast-hood, becoming bestially violent, and then they'll kill us all." Prendick shares this point of view from the moment he learns of the Beast Folks' origins, "urgent to know" why they did not simply fall upon Moreau and Montgomery to "rend" them, accepting the two men's' fear of the "taste of blood"—even referring to its projected result as inevitable. However, this fear of reversion, as they imagine it, does not correspond at all to the actual reversion, which occurs very late in the story, to all the Beast Folk at the same time, and does not otherwise appear. Moreau, Montgomery, and Prendick are all completely confused on this point.

The Leopard Man's story is practically defined by this misunderstanding. When he stalks Prendick, it indicates his urge to hunt, and just before he attacks him, he approaches on all fours, that is to say (in science-land), quadrupedally. It certainly looks like "the beast will out," especially given Montgomery's talk of it the next day.

However, if this were about reversion, then the Law would become meaningless, entirely out of his field of reference, and that's not what happens, not when more information becomes available. When he's exposed for breaking the Law and is chased down, he adopts more and more leopard-like behavior, beginning when he springs at Moreau and culminating in his quadrupedal posture in the final stages of the hunt—but is he reverting, or panicking? All the knowledge we have about cognition and religion describes his behavior as a result of social and corresponding psychological stress—described even to the letter in Paul Gilbert's *Human Nature and Suffering* (1989). No wonder Prendick most strongly perceives "the fact of his humanity" at this moment, which may be paraphrased as understanding the terror the Leopard Man feels, how it arises from the concept of damnation and from accurately understanding what Moreau will do to him. I am left with the interpretation that he goes to four legs in the same way that an utterly terrified person might lose control over his or her bowels or may be unable to speak coherently. It has nothing to do with reverting to animal form or with having tasted blood. This also casts retrospective light on his behavior in his introductory scene, tagging it as rebellion, "acting out," not reversion.

Moreau's and Montgomery's exceptionalism is a rotten thing, with its caricature of both "human" and "animal." It allows the persecutor to patronize the victim, to experience contempt for his or her "animal-like" writhings, to write off his or her cries as "mere animal noise," to sanitize his power to inflict pain and to enforce obedience. As Prendick did not himself manage to conceive during their debate, the elevated Man is not a paragon of spiritual virtue having tea with Spencer's angels, but too easily an oppressor and torturer given license from empathy.

Many Beast Folk display the potential for excited violence. They are definitely ready for it during the chase after the Leopard Man, and the Hyena-Swine takes the chance to sink his teeth into his former friend's body. After Moreau is killed, two Swine Men and a couple of others chew up the Puma Woman's body and behave in a disorganized way. Whether these actions are inhuman bestiality emerging, or similar to human mob violence that happens to be equipped with bigger and sharper teeth, is left unresolved in the text.

The character to watch, though, is M'Ling. Don't let his submissive demeanor in Chapter 3 fool you; he's mainly made from a *bear*, after all, and is described as one of the most physically formidable Beast Men in the story. His personal arc is all about violence, as he's built up as the slow burn candidate for bestial revolt. He has been "taken in" by a human, "treated as one of us," and then abused with insults and blows, on and off by Montgomery, and especially with his humiliation on the *Ipecacuanha*. From there he goes from licking the blood from his hands after skinning the rabbit (Chapter 16), shifting his weapons preference from hatchet (Chapter 16) to teeth (Chapters 17–18), relishing the fight with the Swine Men, and taking an unseemly interest in the body of the one he kills (Chapter 17).

If this were a story about the unleashed frenzy of barely tamed beasts, the humiliation M'Ling suffers in Chapter 3 would be a fine starting point to proceed through this sequence to the eventual rending of Montgomery limb from limb. But the revolt never happens! Sinister scare-hints and all, he staunchly remains Montgomery's Man Friday until he is killed. Reflecting on the entire arc, I am left with the portrait not of dumb-brute submission, but rather a remarkable display of ambition to be a person and of personal control, to manage aggression rather than either suppressing it or producing it whenever provoked, all tragically connected to submitting to Montgomery as less than an equal partner in friendship. I suppose he succeeds—after all, he doesn't die "like an animal" but as a person well might, in a senseless, drunken squabble.

M'Ling is the least faithfully depicted character in the films: always physically unimposing, submissive to the point of cringing, and Moreau's lackey rather than Montgomery's friend. They all devote much imagery in his clothes and demeanor to the distinction between him as house-native rather than field-native like the others.

In *Island of Lost Souls* (1932, Tetsu Komai), he is made from a dog, and called a "faithful dog" in reference to his death at the hands of the rebelling Beast Folk, as he tries to defend Moreau. Although his animal origin isn't specified in the later films, they add their own twists toward the end. In *The Island of Doctor Moreau* (1977, Nick Cravat), he resents his agonized reprogramming enough to deliver Montgomery's body to the other Beast Folk, to prompt their rebellion. In *The Island of Doctor Moreau* (1996, Marco Hofschneider), he is exceptionally wimpy and apparently the most eager to please among the pampered house-servant Beast Folk, and he becomes assertive and vengeful toward Hyena after he murders Moreau. It's impossible to explain why or how he prompts the latter to suicide, one of several impenetrable features of the film in its final quarter.

The Hyena-Swine also tastes blood and also does no such thing as "revert to the beast." We don't even meet him until after the fact, and he's arguably the most sophisticated of all the Beast Men, far from being half-reverted to a hyena-pig. He covers for his transgression by genuflecting piously before Moreau and by participating most vengefully in the ensuing hunt. Later in the story, he only starts seeking to kill Prendick after Prendick has tried to organize the Beast Folk to kill him, and failing that, "[a]gain and again . . . tried to come upon him unawares" (p. 95). He never attacks other Beast Folk, and he kills—wait for it—*one* more rabbit, throughout the ten months after Moreau dies. Not exactly a frenzied rampage.

When the Hyena-Swine reverts to the point that he's willing to attack Prendick despite the latter's pistol, he has done so right on schedule with the rest of the Beast Folk who have not tasted blood. Not only did the tasting of blood not accelerate reversion for him, the extent and process of reversion occurred entirely in its absence for the others. Nor did his murderous intention toward Prendick arise from his reversion; it was already present in force for several reasons, nor, before his reversion, does he act upon it spontaneously.

I'm forced to conclude that Moreau's and Montgomery's conviction about tasting blood was mistaken from the outset, and also that literal reversion of the techniques does not occur even by a scintilla prior to Chapter 21. Chalk up another "poor bastards" instance for the Beast Folk; they probably could have enjoyed yummy rabbit dinners with no erupting-beast consequences at all. All along, it was solely a social issue for Beast Folk like the Leopard Man and the Hyena-Swine: for the former, a covert and guilty sin, and for the latter, the realization that the Law's authority is arbitrarily enforced and contains no intrinsic morality. Their misbehavior isn't about reversion: it's about the human characters' failure to recognize emotion and thought, expressed as rebellion, specifically because they write off its manifestation as reversion to what they see as a sub-human status. That's why Moreau's notion of social governance is best described as "the beatings will stop when morale improves," and why Prendick never, for one minute, considers simply talking honestly with the Hyena-Swine.

When reversion is present in the films, it never appears as a timed and collective event across the Beast Folk, but as an individual slippage, usually continuous and ongoing, featuring a variable mix of behavior and physical form. Lota's hands regrow cat-like claws in *Island of Lost Souls* (1977), prompting Moreau's fury, but she shows no personal savagery until the end, and then it's hard to tell whether this was part of a reversion or whether she was always capable of it if her love interest were threatened. No other character displays this physical regression, unless one infers that it's the general reason that the Beast Folk are especially grotesque during the crowd scene at the end.

The Beast Folk in *The Twilight People* (1972) are always in the grip of sudden furies and passions, so their rampage at the end is no more than an opportunity to direct them, rather than a change in their condition. Similarly, in *The Island of Doctor Moreau* (1977), the Beast Folk exhibit constant hostile and random behavior, which the Sayer of the Law can barely manage; again, it's not a retrogression in Moreau's technique but the general condition of the Beast Folk to be in a semi-berserk state, which only needs a provocation to become a riot. *Doctor Moreau's House of Pain* (2004) ramps this condition up to maximum with a wide range of berserk, impulsive, sexualized, and gory actions, with the same implication regarding the Beast Folks' state of mind.

The mental and emotional version of this kind of regression is especially well dramatized in the 1977 film, when Moreau tries to "animalize" Braddock through reversing his techniques. He reinforces his intended result verbally, describing the changes he expects to see: how Braddock must now be thinking in images, hot-and-cold, hunger, sudden rage, implying that beasts have no categorical cognition and no regulatory or modulatory processing of their emotions. Braddock resists appropriately by maintaining verbal acuity and memory, implying that a special human access to these qualities is the same thing as *keeping control*—that animals have no memory or concept-based mental constructs and run around all the time *out of control*, precisely as the Bull Man did, shouting "Animal! Proud!"

The issue is most scattershot in *The Island of Doctor Moreau* (1996), in which the Beast Folk display a wide range of overt and covert violence, and in which Moreau bizarrely insists that they are "harmonious" and "incapable of harm"—despite using medication and remote-controlled agony to control them. More than one seems to relish hurting others and getting away with it, most evidently Azazello, but similar to Lota, only Aissa undergoes a physical transformation "back to the beast."

Regression in the films completely accords with and reinforces the Man/Beast divide. When physical changes are involved, the visual techniques borrow directly from the vampire and werewolf cinema, especially growing claws and fangs—features that are definitely absent in the novel, in which the Beast Folk's teeth are not amenable to surgery and always look like their species of origin.

Sudden emotion and violent action are an essential subset of human exceptionalist thinking and its amazing capacity for projection. The scene I described for the 1977 film captures it perfectly: apparently nothing says "beast" and codes it away from humanity like—let's see, striking the bars of one's confinement in frustration, losing one's temper when being tortured, having trouble speaking clearly when drugged, having trouble remembering a word when under extreme stress, and using any available weapon when attacked. An observer might be for-given for thinking that this and any similar film got it backwards and mistakenly put in human violence where the "beast" one was indicated.

It's quite a job to help students see that nonhumans do not live in a moment-to-moment haze, punctuated by bizarre upwellings of inexplicable and randomly directed violent urges. This model is embedded so deeply in our terms like "ani-malistic" and "bestial" that one has to exert some effort to throw it off. What goes with it, too, is the model of one's thinking self as necessarily calm, collected, and mild.

Brain and Behavior

Beginning with the physiology, one difficulty arises with identifying the vari-ables. The most straightforward data concern the production of violent action and brain activity, because both can be measured, but they do not tell us much about the internal experience of anger or impulse. Studies that combine the "I" agency concept described in Chapter 6 with aggressive responses might get at this conundrum, though, and quite a bit of insight comes from monitoring physiologi-cal indicators appropriate to the species, like urine or fecal chemistry in rodents.

That said, we know a lot about brain activity during agonistic situations. How does it work? Before you say "testosterone," let me stop you. In the brain, the most relevant agents are neither precisely nervous impulses nor hormones, but a different kind of protein that is conducted by neural tissue, collectively called neurohormones. Neurohormones have remarkably diverse functions, but in ref-erence to brain function and aggression in mammals, serotonin is the primary inhibitor and vasopressin is the primary facilitator. They're made and conducted in extremely specific paths through the brain during agonistic (conflict) situa-tions, leading to equally specific areas of the cortex becoming electrically active (Figure 9.1).

The complexity of the system gets dizzying from there, but that's enough for my purpose. Even the basics of these processes eliminate a lot of mythology quickly, specifically, that there is no single switch to flip that jacks a brain, or a person, into a beast-mode of uncontrolled mayhem.

There is no demonic single "angry substance" that infuses your brain and makes it stupid and mean. Although testosterone and its derivatives are a hormonal component of the vasopressin brain pathway, it does not directly cause aggres-sion in a dose-response manner. Also, as in most physiology, the receiver matters as much as the signal, so the cortical neurons' receptors are a dynamic property in the whole response as well. As for the inhibition, it's even more nuanced: you can see from the diagram that the serotonin response is a highly processed event with important connections to your memory, meaning that it's individually spe-cific, both to that person and in that moment. Serotonin is probably the more

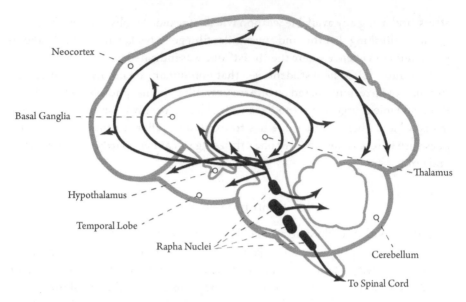

Figure 9.1 Serotonin and vasopressin in action.

important substance in that instant, too, because the inhibitory regulation is constant. Crudely speaking, that means that you are always on "go"—that's the vasopressin, among other things—but you keep the serotonin brakes pressed most of the time, so producing the behavior is more about the brake letting up than about the "go" revving up. Even this system is made fuzzy by the effects of other neurohormones and substances, too, including gamma-amino butyric acid (GABA), noradrenaline, and nitric oxide.

Nor is there a single "angry spot" in the brain devoted to that particular response, and most especially, there is anatomically no such thing as the "mammal brain" or the "reptile brain." This particular bit of popular phrasing would do well to evaporate as soon as possible, as follows.

- All vertebrate brains have the same three parts and subdivisions, and they're all hard at work, regardless of their various proportions in terms of volume.
  - Specifically, the pallial portion of the forebrain is always present, and although the mammalian version has six distinctive layers—the neocortex—that's an elaboration, not an added piece with a completely novel function.
- The basal ganglia of the midbrain, long saddled with the reptile term, is implicated in behavior-switching, or what you like to call "choice" (so at least you can stop calling it names).
- Similarly, anger and aggressive actions do not reside only in the mid- or hindbrain, but are neurologically tracked through a great deal of the whole organ,

especially the famous frontal lobe of the cerebral cortex, often touted by non-biologists as the "rational" part. This allegedly rational part lights up when you're hitting someone, apparently.

• Vasopressin, the neurohormone associated with aggressive response, is a forebrain product. Serotonin, the neurohormone associated with inhibiting that response, is a hindbrain product.

The tempting imagery of the triune brain was popularized during the 1970s by Paul MacLean and was reinforced in his 1990 book *The Triune Brain in Evolution*; from there it has become a pseudoscientific staple in self-help and business-school texts. It began as and remains counterfactual exceptionalism. Again, there isn't a thinking part that suppresses a stupid or reflexive part. When a behavior is fast and intense, that doesn't mean stupid or "out of control." When you get mad or do something harsh and violent, or both, you're thinking differently from when you write a sonnet, and you're certainly *talking* less, but you're thinking just as much.

The list goes on, eliminating story after story. For example, there are definite sex differences in the physiology of aggression, but not in the simplistic sense of "males more, females less." As far as the species studied are concerned, mostly humans and rats, female aggression is less obvious (to us) and less frequent, but no less available and no less dangerous—if anything, to the contrary.

That observation suggests that superficially similar aggressive acts can well be produced under entirely different brain states, or that there isn't a single identifiable aggression. Further brain-and-behavior work distinguishes between two types: controlled/instrumental or reactive/impulsive, which is associated with less prefrontal cortical activity. When taken to clinical observations in humans, this idea is confirmed in that post-traumatic stress disorder, intermittent explosive disorder, irritable aggression, and depression-linked aggression all include heightened autonomic responses, like faster heart rate, higher blood pressure, sweating, and dilated pupils, but conduct disorder and antisocial personality disorder do not.

Social and Interactive Analysis

The brain work crucially reveals that aggression neurologically occurs in the same multipart complex that manages most social interaction. Social creatures come into personal conflict frequently, within their own species and group, and a whole suite of behaviors is centered upon inflicting harm or threatening to do it. This is not a disruption or a violation of social activity but is instead embedded in it, resulting from it, and often elaborated in exactly those terms. These are common and recognizable actions—basically, do you or do you not attempt to kill or seriously injure the person next to you—with highly species-specific circumstances and standards for when, about what, with how much provocation, and in doing it, how much damage is delivered.

The long-held claims that social violence is a function of territoriality and dominance hierarchies have not been borne out by research. Perceived reputation and organized or semi-organized group dynamics are apparently the topic at hand instead. That means that studying aggression as brain response or as behavior will not be comprehensible, or even sensibly analyzed, outside the larger social context, meaning the potential response from organized third parties.

Sometimes the topic of a dispute or violent action, "about what," is specific, common, or predictable, and very consequential in evolutionary terms, as with large social contests that play heavily into who gets to mate, and with whom; or collective organized actions such as raids. The degree of this kind of violence (and exactly who its participants and targets are) seems to be ritualized and understood by everyone delivering it. In the case of mating contests, the degree of harm is usually bruising but not too dangerous, with some exceptions, whereas raids include sadism and atrocity. This degree—whatever it happens to be in a given case—is also apparently socially supported (i.e., acceptable), and its outcome provides data for others' social actions. For these organized or collectively supported actions, these observations suggest that strong selection has resulted in a narrow and specific range for each species.

Opportunistic, less-organized violence is harder to assess, both logistically and also relative to our own strong reaction to such things as bullying, humiliation, assault, and the nuances of provocation and the degree of response. These and other actions include the grim realities of sudden yet clearly targeted deadly or very injurious violence: bar fights, murder arising from an argument over any disputed thing, bashing and often deadly harassment, murder or abuse of kin who are not behaving as desired, murder or assault toward a perceived romantic rival, and abuse of one's mate or offspring. Research into these actions repeatedly shows that one's social positioning and reputation—what others are saying—play heavily into its instigation, suggesting that they are not counter-social or anti-social, but are fully embedded in the ultrasocial matrix. That does not stop organized third-party response from isolating and punishing the perpetrators, but it also does not stop such parties from protecting them either—again, it's a matter of social positioning and community identity. That's why selection upon the limits of violence and social norms are not the same thing.

At the personal level, such violence reminds me of the odd observation, in my experience, that a person sometimes insists that another person *agree* to fight, then responds to that agreement as a provocation.

Research at this level of limited, relatively individual agonism in an evolutionary context suggests that selection favors a modulated, personal, and wide-ranging response, rather than a switch-flip into a less thinking state. One major branch of analysis along these lines relies on mathematical game theory, looking for algorithms that our brains, or rather, ourselves, we, might use to negotiate our way through a tricky and consequential landscape of socializing and physical risk. The most famous programming model specifies Hawk (violent) and Dove (peaceful), and examines the outcomes based on their proportions. Typically, a "third way" option is included, one of many with names like Bourgeois and Retaliator, which modulates violence based on circumstances. The idea is to see which tactics remain in place or become more common, depending on the starting proportions, and which other tactics are included; the presumption is that these dynamics may operate as selective factors on the behavior of real creatures.

In reading about this, it's easy to get confused by the colorful names and think they are personality indicators. The modeled terms are not intended to represent individuals but rather behaviors, and therefore the outcomes can represent either the proportions of individuals with fixed behaviors or the proportions of behaviors produced by any single individual, usually the latter when taken to the field for further testing. It definitely does not mean that every given real person is designated to be a "Hawk," "Dove," or whichever other term is included.

The primary insight from this body of work is consistent with the brain work: whatever combinations are employed, the most successful ones (they call it "stable") are either adjustable, shifting among tactics based on circumstances, or contingent, shifting among tactics on a percentage basis. When the model is taken to observational work with real animals, and if the researcher is diligent in using wild or semi-natural circumstances, usually the results correspond to one or another of the "third way" options predicted by the model. Species differ greatly in the details, in terms of the exact proportions of the behaviors, the subtlety of the potential combinations, and in whether the shifts are adjusted or contingent, but overall the model seems to be borne out. The general point to bring to the Man/Beast divide is that human animals aren't angels (Man) because they can cooperate well, nor are nonhuman animals demons (Beast) because they can be harsh and cruel, nor is either statement made true if you flip the subjects. We as well as other creatures shift and adjust our potential violence relative to what one another is doing, or to put it most precisely, the human way to cope with and to deliver harm to one another—including not doing it—is one of the many animal ways.

A lot of these writings, especially when popularized, are rife with the usual idealization of selection, whether right in the text or too easily inferred from their phrasing. It's well to remember that a species isn't "fitted" to its history or environment, or "perfectly adapted" because it's got a selective history, but is a rather messy outcome of a lot of colliding variables. Therefore one does not look for pure or perfect bundles of tactics. One oddly persistent concept is the evolutionarily stable strategy, which is nothing more than the concept of frequency-independent selection applied to an array of behavioral options. The persistence is odd because plenty of selective processes are frequency-dependent, yielding jumping-around proportions of variant traits, generation after generation, rather than a single ever-increasing trend toward one of them (in this case, a specific tactical array). Nothing about selection on behavior suggests that it should uniquely be density-independent, and one should be quite wary of the idea that any species exhibits a cleanly designed, all-bases-covered, and completely stable array of behaviors due to some magical interpretation of selection.

How does any or all of this relate to the moral experience described in Chapter 7? It's incredibly difficult to discuss because any action is subject to collective approval or disapproval, raising profound stress regarding one's willingness to submit to or fight against the group judgment, and raises the question of whom that judgment genuinely benefits. Many secondary issues arise as well, including personal guilt or shame, and the role of designated criminality for specific actions or persons. Criminality is a social designation, not an identifiable group of actions, so an individual in an ultrasocial species is always at risk of policing—personal choices are always relevant to the others in the community, and are always subject to organized response, itself under no guarantee to be fair or immune to others' power-games. One might hope for an interdisciplinary attempt to discuss these matters, much in the way that the study of brain-and-behavior has managed to close several gaps, but nothing like that regarding sociology and the psychobiology of aggression seems forthcoming soon.

My point in Chapter 7 regarding an expected failure rate for any behavior, no matter how strongly selected, adds a further layer of complexity in the management of inflicted harm and threats to do so. How might we tell the difference between a violent strike that represents a successful tactical decision honed by selection and one that represents a cognitive and social mistake? This is a modulated response in contingent circumstances, which is to say the potential for ultrasocial judgment with extreme consequences, in which the long-term fallout is impossible to predict.

An expected intrinsic failure rate is only the beginning of how things get murkier and more difficult to assess in the existing research, both observational and theoretical. There's also an uneasy and variably articulated division between ordinary, typical, socially unremarkable violence, which may sometimes be quite extreme, and over-the-line, disapproved violence, which brings social condemnation and collective punishment. Other important distinctions include personal or one-on-one fighting, assault upon a targeted and usually less able individual, sudden and violent group activity toward one target or toward another group, and organized and repeated group activity. It seems likely that in an ultrasocial context, these aren't merely variations of a single behavior, but experienced and modulated as very different things. Studying this range neurologically is very hard or impossible given our current technology, so most of the neurological research concerns one-on-one confrontations—unfortunately missing out on the typical complex social context that matters so much to us personally, and which may have exerted the strongest selective pressure.

Insight into such subtleties would make the fight with the Swine Men in Chapter 17 more interpretable. It could be that they have gone silly with violence and are now altogether descended into the Beast, but it's also true that humans have been known to mutilate bodies and otherwise behave atrociously in similar circumstances.

No evidence suggests that humans manage personal violence better than other animals, and some researchers think we're markedly worse at it than most. The argument goes that since we're not very well armed compared to most mammals, the risk of mistaken aggression is lower, and the ultimate selective result is that we're tone-deaf when it comes to spotting potential violence or resolving escalation. It's an application of the "primeval man trapped in the modern world" concept, suggesting that lethal technology has made inadvertent killers out of our otherwise posturing or bluffing behaviors. However, this idea has not yet been assessed by comparative or physiological work.

It's also hard to tell how any of this research relates to the category of pathology, and how to assess the difference between a very extreme violent action and an insanely violent action. It doesn't seem to be easily designated simply by the degree of harm involved. A whole branch of brain research is dedicated to the imaging and brain architecture of psychopathy, and it points to a big difference between the two, which put simply, means that ordinary people get violent because they care very much, whereas psychopaths do not. But it doesn't mean that ordinary people can't produce the same levels of violence. The distinction seems relevant to the point, though, because psychopathy is about as close to "the other" one can find in reality, and therefore the stereotyped Beast, conceived as across the divide, is sometimes coded as a psychopath.

Discussing the issue runs into multiple assumptions and pitfalls among scientists and non-scientists alike. For instance, it's difficult to expel phraseology concerning blood, for some reason. Social anger and aggression aren't prompted by the sight or smell of blood, but it triggers social and judgmental responses that get projected back into the violent actions.

We also have trouble assessing nonhuman animals on their own cognitive and social terms. When nonhuman animals attack members of their own species and do more damage than *we* think is appropriate, then we misinterpret it as unmanaged anger or psychosis. This extremely tempting perception facilitates a number of other denial tricks:

- Eliding discussion of our own quite considerable capacity for violence
- Assigning to nonhumans poor management of aggression and misunderstanding of cues

- Avoiding the idea that selection may have favored a standard or average level of accepted harm of certain kinds
- Misreading another species' ordinary ability to do damage as our version of excessive force.

The same projection we bring to brain studies is rife in this work, too, our inaccurate and quite unfair notion that nonhuman animals are continually in the grip of hostile urges. It's a special kind of anthropomorphic projection, in the sense of ignoring whatever it is experiencing and feeling in favor of a construct made from what its actions would imply from a human.

When humans and nonhumans socially overlap, as with dogs and horses, social aggression produced by a nonhuman may be regarded as criminal by the humans, or at best, as lack of competence, resulting in criminally unacceptable action. Arguably, when a creature attacks you socially, then you may have pissed it off in some fashion, perhaps with bad judgment on its part but also perhaps not. These events threaten our sense of control and safety, and our sense of the rightness and wrongness of violent reactions snaps immediately into social human terms, with the accompanying position that this creature has no business judging a human or acting upon that judgment. Assessing these issues is so loaded that I don't think we're even close to understanding the mechanisms.

Crucially, we confound aggression and predation, because the latter is plain violent and injurious, and that's if you're lucky. When a creature tries to eat you, I can't say I know what it feels, but no research suggests that it's socially mad at you despite being willing to kill you. But we interpret its actions in social terms because we interpret everything that way, so it must hate us, or be angry at us, and if we "didn't do anything," then it must be sociopathic or psychotic. Blood is overrated here, too; predators may well use it as a tracking cue, especially in water, but it does not excite them in a special way or make them go berserk, not even sharks. The so-called "shark feeding frenzy" is nothing more than a lot of sharks eating a lot of fish—it's full of scary teeth, threat displays and bites among the sharks, and very gory regarding the prey, but nothing the sharks do indicates they've gone bananas.

I don't know who first claimed that herbivorous creatures are docile, peaceful, and social whereas carnivores are aggressive and vicious, but flatly, this is absurd and is absent from any serious observation or analysis of behavior. Aggression is a feature within sociality, not opposed to it or some failed attempt at it, and it has nothing to do with diet. Herbivorous social animals display every detail of the aggressive and dangerous activity I've been describing, in addition to the cooperative and reciprocal ones, and carnivorous social ones display every detail of the cooperative and reciprocal activity, in addition to the aggressive ones. This unfortunate fantasy has become embedded in our muddled discussion of human ancestry, as people jockey to tell our "story" as either grain-munching consensus-builders corrupted by civilization or as romantically ruthless blood-stained murderers seizing the future. You may recognize the nineteenth-century views in action, still, and yet again.

Deaths of Moreau and Montgomery

No, the Beast Folk do not turn on Moreau in a frenzy of reverted bestial rage. The Puma Woman strikes back as an individual, very likely a thinking one, and when pursued and shot, batters him to death with her chains. The Swine Men, the Ocelot Man whom Montgomery shoots, and the "feral monster" Prendick soon shoots were presumably mutilating and partially eating the Panther Woman, but they did not attack Moreau or interfere with his body, because her body has been partially eaten but his wounds are described only as those inflicted by her chains. Moreau's death cannot be described as due to the resurgence of "the Beast."

Montgomery dies in a drunken brawl, but the precise events are unseen. I confess to being amused at the thought of Montgomery leading a tipsy chorus of "Confound old Prendick," but imagining when it goes sour is depressing. Prendick hears the singing die down, then a violent dispute arises, with cries of "More! More!" suggesting that the liquor has run dry, to the carousers' anger. Prendick arrives too late: M'Ling is dead, his neck bitten open, still clutching the broken bottle; a Wolf Brute is too injured to live (Prendick shoots him) and a Bull Man is already dead; and Montgomery and the Sayer of the Law (here, the Grey Man or Grey Thing) are lying together, the former strangled by the latter and dying soon afterward.

All right, so the liquor ran out and the party turned into a brawl, which is easy enough to understand. Who exactly killed each Beast Person is hard to say. M'Ling may be easiest for speculation, supposing that he was defending Montgomery and a sharp-toothed Beast Man got him, probably the Wolf Brute, judging by the injury. Montgomery fired only twice, so supposing both shots were effective, that leaves one of the assailants unaccounted for, unless M'Ling took down the Bull Man first. Or maybe the whole thing is opaque to forensics, considering that a whole bunch of Beast Men were struggling together before Prendick arrived and dispersed them, so nearly anyone may have killed anyone, with the exception of the one certain case, that of Montgomery.

Or is it? Why would the Sayer of Law (his identity is confirmed in Chapter 20), of all people, kill Montgomery? It can't possibly be about the dreaded reversion to "the Beast." The Sayer of the Law showed no signs of reversion, and in fact none of the Beast Folk has begun to revert at this point in the story. Nor was he identified previously, by name or as the Grey Man, as one of the revelers and brawlers. He had even just been convinced by Prendick to uphold the Law.

I'm tempted to infer that the Sayer of the Law is acting as a religion-driven temperance extremist, driven past patience to punish not only a sinner, but a tempter and corrupter. However, this reading is not explicitly justified in the text: no part of the Law recited earlier mentions alcohol, and it doesn't seem to fit the pattern of what the Law does say. The scene is visually murky and seems to match its thematic content being murky, too. The only certainty is that neither carnivory nor reversion has anything to do with it, and I'm unpleasantly left with the all too human image of drunkards shouting "More! More!" rather than nonhuman predators hunting and rending.

The Stubborn Beast Flesh

Let's take a look at Prendick during the next phase of the story, the only time when he refers to the Beast Folk as Beast Monsters. It lasts ten months (!) of his life among the Beast Folk. Briefly, just at the start, Prendick experiences a moment of clarity:

> Were they peering at me already out of the green masses of ferns and palms over yonder—watching until I came within their spring? Were they plotting against me? What was the Hyena-Swine telling them? My imagination was running away with me into a mass of unsubstantiated fears. (*The Island of Doctor Moreau*, p. 91)

How truly he speaks. But instead of recognizing his own projection, even as he describes it perfectly, he descends further into paranoia (his own word). In Chapter 8, I described the Beast Folks' dramatically sensible assessment of their new religious life, with no trace of savagery or incivility, or for that matter, "plotting against" him:

> We love the Law and will keep it, but there is no pain, no Master, no Whips ever again. (Ibid., p. 93)

But Prendick's reaction to this exact statement is nothing less than hysterical. He perceives it as a direct threat and instructs the Dog Man to be ready to attack the Hyena-Swine at Prendick's signal. Here is the context:

> I still went among them in safety, because no jolt in the downward slide had released the increasing change of explosive animalism that ousted the human day by day. But I began to fear that soon the shock must come. (Ibid., p. 97)

He is so terrified of this "shock" that he transforms it into a fact, not realizing that he's literally inventing "the explosive charge of animalism." He doesn't give the

Beast Folk any credit for decency even when they shelter and feed him, which he apparently continues to rely upon throughout this period. And how does he act among them?

> In the retrospect it is strange to remember how soon I fell in with these monsters' ways, and gained my confidence again. I had my quarrels with them of course, and could show some of their teeth-marks still; but they soon gained a wholesome respect for my trick of throwing stones and for the bite of my hatchet. And my Saint-Bernard-man's loyalty was of infinite service to me. I found their simple scale of honour was based mainly on the capacity for inflicting trenchant wounds. Indeed, I may say—without vanity, I hope—that I held something like pre-eminence among them. One or two, whom in a rare access of high spirits I had scarred rather badly, bore me a grudge; but it vented itself chiefly behind my back, and at a safe distance from my missiles, in grimaces. (Ibid., p. 95)

Really, Prendick? A pecking order based on inflicted damage was conspicuously absent in Chapters 11–12 when he met the Beast Folk in their homes, in his general description of their ways in Chapter 15, and in their approach toward him at the end of Chapter 20. Already self-acknowledging himself as paranoid, it looks as if he moves in and simply starts knocking them around. There's no way to confirm this outside his own narrative, except that he completely ignores his own observation that he lives among them in safety. Even more charming, in the grip of his fear of what he imagines, he plots to exterminate them all gratuitously.

In the longer term, Prendick is proved wrong. Ten months go by, with no House of Pain and no Moreau, and he states outright that the Beast Folk follow the Law "with decorum." He reports no fights aside from his own, no murders, no riots, and no devourings. A look at Prendick at this point isn't edifying: he's a wild-eyed survivalist with his visions of imminent attack, his weapons, and his dog, jumping and starting at the mundane and apparently staid activities of ordinary villagers, and given the most unsympathetic reading, even prone to hacking at people with his hatchet. When he raves to them about the fellow whose only crime is to prefer to live by himself, they refuse to form a lynch mob for him, so he then devotes himself to trying to kill this guy. I wonder if they thought of him as a socially disordered bit of a burden—it's hardly likely that he did his bit at farming, for instance.

When the Beast Folk do revert, it apparently begins abruptly and proceeds rapidly, and the result is very different from Prendick's expected bloodbath. Their reversion is sad and quiet, not violent, more like Alzheimer's disease than anything else. The worst he can come up with is that they scream a lot at night and have lots of non-monogamous public sex, bad enough to a nice boy from England I suppose, but they make no effort, individual or concerted, to hunt him down.

The Hyena-Swine's final scene bears special attention. Let the record show that he has killed the Dog Man and is eating him, but has not in fact hunted down Prendick, although I see no reason to doubt that he might do that next. The two have been attempting to kill one another by ambush for some time, which began long before anyone started reverting, so if the Hyena-Swine attacks Prendick, there's no reason to go reaching for the reverted Beast for a motive: he already hates Prendick's guts for several pretty good reasons. The one element introduced by his reversion is that he now poses a *lesser* danger to Prendick than he did before, and it proves his downfall, because he can no longer perceive the gun as a threat. Prendick totally misses this point and interprets it as if the reversion motivated the attack and as if it heralds similarly motivated attacks from the others.

No wonder the films simply abandon fidelity to the novel's plot for the second half. I can just see the director or screenwriter flipping the pages, baffled—"Hold on a minute, where's my vengeful uprising?"

The "don't meddle" story says, "You can't be a man, you're a beast," preserving exceptionalism, with all sorts of versions and consequences for how wrong it is to try. To reconcile this idea with even a modicum of faithfulness to the novel results in intractable tension between critiquing the exceptionalism and milking it for the best standard horror-action. Plot trouble results: making them more or "really" human amps up the injustice component, especially if they're beaten or exploited, but that conflicts directly with making them in-no-way-human, instincts-take-over dangerous, and going with the rampage.

The justified rebellion and the crazed-beast rampage don't mesh well as concepts. When the film tries for empathy, it directly contradicts "don't meddle," so it gets diverted to "it must go wrong somehow," undermining the empathy. When it relies instead on "the taste of blood," it makes no sense relative to the Beast Folks rebelling in a rational understanding that they're being screwed, and turns it into a gore fest and a rampage, which in turns diminishes the injustice component of "don't meddle."

Each film chooses one or the other as the final plot device. The interpretation of M'Ling always indicates how a given film wants to swing it, also acting as a measure for how coherent the story turns out to be. *Island of Lost Souls* (1932) leans toward the rampage-carnage option (Lota aside), and M'Ling is the only Beast Man to defend Moreau against his just desserts, but in doing so is just as animalistic as the others, strictly the "faithful dog," as Montgomery disparagingly calls him. *Terror Is a Man* (1959) switches from the rampage/carnage to the empathy quite sharply. *The Twilight People* (1973) is a fine exploitation B-flick, which knows exactly what it's doing and goes with the rampage/carnage all the way through.

The films titled after the novel each go for a compromise and perfectly demonstrate that it's impossible. *The Island of Doctor Moreau* (1977) illustrates the point through the confusing behavior of two characters. The Sayer of Law encourages the attack on Moreau, but tries to stop the destruction of the compound and the opening of the predators' cages, and he and the others are killed by those animals, which means precisely nothing, thematically. M'Ling bounces back and forth between the two concepts like a ping-pong ball. He begins as the culprit at sucking up drink and has to be reprogrammed early in the story; then he informs the other Beast Men against Moreau, but finally he ends up defending Braddock and Maria against the others' rebellion, even dying in classic sidekick style to ensure their escape. *The Island of Doctor Moreau* (1996) goes all the way with the rampage/carnage, right when it might have profited most not to, subverting its own buildup that the Beast Folk worship and suffer in sympathetic ways and are effectively a captive brainwashed population. Its plot breaks down entirely when the two most thoughtful ones, Hyena and Aissa, become ridiculously random when they "go animal," which immediately results in their own deaths, as well as Azazello's, who seems to have retained some sense until he makes the mistake of going anywhere near Hyena.

I am convinced that this film is so curiously unsuccessful in terms of story-enjoyment not because it simply fails to be good, but because it never manages to put several good ideas into a sensible set of relationships and caused actions. Again, M'Ling is the touchpoint for the confusion, initially displayed as the most obedient and wimpy of Moreau's house-creatures; then he's the one, clutching his precious book of poetry, who blows up the villa to help stop Hyena and seems somehow to prompt him to commit what appears to be remorseful suicide in the fire—which makes no sense insofar as Moreau has been tagged as functionally insane. Furthermore, this exact climax is then ignored in the contradictory final scenes, the first of which goes with the Sayer's line "Perhaps four [legs] are better anyway," and the second of which goes with Douglas's narration, which fears human violence as an emergence of bestial urges.

Dr. Moreau's House of Pain (2004) plays with both concepts but surprisingly turns toward the empathy option, albeit with plenty of disgusting acts and considerable gore along the way. The two rampaging Beast Folk are motivated by understandable personal concerns, and although they are spectacularly violent due to their physical capabilities, in terms of cruelty they take second place to the human characters and die with at least a little sympathy on their side.

The Horror

The reverted Beast Folks' most terrible moment for Prendick comes when a couple of them begin to eat the corpses in the boat that has washed up on the beach, one of which may well be Captain Davis. May I say in aside, I experience no reader's sympathy for this person or his remains—as I see it, his death by thirst, just as he'd planned for Prendick, is nothing but poetic justice, and the most useful thing that guy ever did was to get eaten by animals. But it provides one crucial difference for Prendick: so far, the instances of Beast Person eating another one have been very few, and now, for the first time, the reverted Beast Folk are eating original human meat. As he launches the dinghy to escape the island, he does not look back to see it. Understandable, certainly—I probably wouldn't either. But there's more to it.

Human Flesh

The story of his adventure begins with him confronted with the grim reality that humans, in extreme circumstances, might kill and eat one another—to which, in fact, he agrees—and he is only saved from meeting this end himself, one way or the other, because the other two men fall overboard. It ends, excepting the final chapter, with the very same kind of boat and the very same eating of corpses he'd almost committed in it, and he refuses to look at it. That refusal means a lot to me, as a reader. Let's take a look at cannibalism.

What is it? An analysis begins with the observation that few if any animals include it in their lives as run-of-the-mill, ongoing predation. The current thought in evolutionary theory is that preying on someone who shares exactly your own profile of parasites is—as the physicians would say—contraindicated. At least in terms of effects on selective outcomes, avoiding cannibalism is a solid instance of hygiene. We are not talking about some bestial behavior that is common for mere animals but which exalted humans avoid through their culture. Yet again, a behavior we like to claim for ourselves and to consider a unique achievement turns out to be something any number of animals do—in this case, almost all of them.[1]

Interesting routine or predictable exceptions do occur, including cannibal tadpoles who avoid eating their own kin, or male rats who kill and eat a female's young litter prior to mating with her. I say "interesting" because young creatures might be carrying fewer parasites and therefore less risk is being run, but also because these instances represent highly specific circumstances of social interaction. Certain other behaviors can be interpreted as cannibalism at least physiologically, as when pregnant female mammals of many species, if food becomes very scarce, resorb the embryos, effectively digesting them through the uterine wall.

Now for the non-routine cases: when animals in a generally non-cannibalistic species do it anyway. In humans, three types stand out.

- Some humans literally prey on others secretly. The ones who've been caught and studied score high in specific brain and behavior anomalies, in the always confusing metrics of sociopathy and psychopathy. The insight seems to be that the individuals literally do not belong to the communities in which they live, and their behavior is best understood as a very broad-ranging social pathology, rather than a single specific act of interest.
- Ritual cannibalism as a social norm is also observed, but here the academic culture breaks down into much strife. There are two camps: one that considers it a rare act and that most of the gaudy reports of "cannibal tribes" are colonialist tale-telling; and one that doesn't. Current evidence suggests the former are probably correct, but every anthropologist takes a side on this one, and I dare not tread there.
- Humans will eat other humans in emergency starvation situations; in this, there is no discernible difference from most other animals. This is the case most relevant to the novel, and fortunately, the least difficult to discuss in analytical terms.

The depiction in the novel is drenched in immediate history. The text explicitly mentions the famous wreck of the French frigate *Méduse* in 1810, after which over a hundred castaways were abandoned on a flat raft; during the fifteen days adrift, many were swept or thrown overboard, but some were eaten, and only ten ultimately survived. Two other cases are unmentioned in the text, but are much more recent and of public interest. The first came with the sinking of the British collier *Euxine* in 1874; in one of the dinghies, James Archer and August Muller killed and, with the other castaways, ate Francis Shufus, surviving twenty days. They admitted to the act upon rescue, as it was widely assumed that this is not a crime given extreme emergency. They faced charges that were never tested in court, as the case was dropped due to jurisdiction problems.

The second came in 1884 with the sinking of the British yacht *Mignonette* and a similar situation in a dinghy, during which Tom Dudley, Edwin Stephens, and Edmund Brooks killed and ate the seventeen-year-old cabin boy, Richard Parker. In both cases, the men drew lots to see who would die. They were able to survive for five weeks until they were rescued. Dudley and Stephens were charged with murder in the first British case for this kind of event, notorious not only for its gruesome content but also for irregular abuses of the court process by the judge and for wide swings of public opinion.

Prendick's experience in the dinghy of the *Lady Vain* is a hair short of identical with the testimony provided by the three survivors of the 1884 wreck. For example, Parker did not lose the draw, but he was helpless, so the others—through signals—agreed to kill him. Also, like Brooks, he initially refused to agree to the casting of lots, but later came around to agree. Montgomery all but asks him directly whether he'd killed and eaten anyone to survive in the boat, and Davis outright calls him a cannibal, adding a little more content to the obnoxious captain's hatred for him.

How does this relate to the Beast People? Urges to eat one another, even suppressing such urges, doesn't seem to be an ordinary part of their lives, and even when breaking the Law, the Leopard Man and the Hyena-Swine go after ordinary rabbits. Such urges may be there, but it takes a lot of social rebellion and confusion to see them arise, as with the Hyena-Swine toward the Leopard Man's body and the several individuals who eat the Puma Woman's body. Even after their reversion, Prendick doesn't mention any such activity, although it can't be ruled out once the wolves and so forth found themselves among llamas and goats.

Consider Prendick carefully now: given his mental state that he's described in such detail, this is the Beast Folk at their Monster worst, their origins unequivocally revealed, their dismissal of human social norms evident—and what they are doing is exactly what he, himself, Edward Prendick, human, had almost done, had even agreed do to, in exactly the same circumstances. Instead of seeing the difference between Man and Beast confirmed, he knows how they feel.

The comfortable membrane between himself and the Beast Folk is not only gone, it's rendered impossible. He can't slot "see how bad they are" into the Beast category, as he would go there with it. His carefully managed articulation of exalted humans in the image of the divine Creator, opposed to smelly slinking beasts, can no longer be upheld simply by strengthening its compartments, and he begins his slide into clinically neurotic cognitive dissonance in the final chapter.

Back Home

The horror is subtler than a mere fear of monsters. When Prendick returns to London, he is not afraid of fellow men *becoming* monstrous carnivores or breaking out in stereotyped bestial violence, but of seeing too clearly what they already unequivocally *are*, in their ordinary day-to-day actions. He does not, for instance, refer to the readers in the library as behaving like wonderful intellectual humans who could suddenly drop their books to turn to rend him in a mob, but rather that their activity of reading is indistinguishable from animals engaged in the hunt. He does not fear that the preacher will leap over his podium in a carnivorous frenzy, but rather that he will find the high-minded phrases all too identifiable as nonsensical "Big Thinks."

He mentions reversion only once, and not as a rebellion, a riot, a war, or a mass attack of any kind. His word is "degradation." He fears it specifically as a possible confirmation of his current insight without recourse for further denial. In other words, he does not fear a physical threat from it, but instead seeks to deny his fuller, unequivocal realization of what he already suspects: that the ordinary experience of being human, for everybody, is indistinguishable from the life-experiences of the Beast Folk, even that their worst imaginable behavior is the same as ours.

I do not credit the widespread description of the novel as symbolizing the purported mass breakdown in civility and sense that would soon result in World War I, implying that the war itself was an inexplicable outburst of inner bestiality, and that the novel presaged or predicted it. I have plenty to criticize about these implied causes or qualities of the war itself, and for the tendency to point to art and literature in the years preceding it as somehow infused with its future essence. For my purposes here, I simply say that Prendick is quite clear about what he sees and what he fears, and that no description of war or mass violence is present. He simply cannot stand seeing what is true: that the people around him are in the Valley, too.

Again and again, he protests that he knows the people around him, engaged in ordinary things, are not animals.

> I know this is an illusion, that these seeming men and women about me
> are indeed men and women, men and women forever, perfectly reason-
> able creatures, full of human desires and tender solicitude, emancipated
> from instinct, and slaves to no fantastic Law—beings altogether differ-
> ent from the Beast Folk. (*The Island of Doctor Moreau*, p. 103)

This pleading is nothing but desperation as his denial erodes and becomes so ineffective that he must flee human contact entirely.

I'll share something important with you: I know what he sees, and so do many of my colleagues and students. When deep in the demands of a biology majors program, especially if it's a full-range curriculum rather than pre-professional training, a student may undergo a "Prendick" experience. It begins with the realization that her knowledge base is more than a set of factoids, but a genuinely different interface with ordinary reality. A friend talks about fat-free food or how his cat loves him, and she sighs inwardly. Someone talks about *a rat* with horror, and the student's gaze flicks to the nearby pigeons, noting the nearby actual health threat. She names each bone and muscle as they eat Thanksgiving dinner, until her offended family makes her stop.

It's a strange experience, which first seems like cold distance and mental dissection, "dehumanization" perhaps, or at least that's how it's interpreted if you're foolish enough to tell people about it. I, for example, have learned not to reveal that I have been idly assessing the morphometric distance measures of people's skulls. So far, though, this experience is the same as for any content-rich training.

Then it hits harder. They unexpectedly discover their X-ray eyes, spotting the articulation of a person's pelvis and the bones of the legs, knowing which muscles are alternately pulling such that balance and momentum are preserved. They see, and I do mean see, that everyone is naked under his or her clothes, without shame or arousal. One partner in a nearby romantic couple threatens another person for little or no provocation, and the term "mate-guarding" comes to mind. Sitting on the urban train, they suddenly realize that anyone around them is a potential friend or close-community member, if only they would share a little communication, and they wonder what social stricture has become so strong that this barest act of contact is discouraged. They snuggle into bed and think of themselves as nice fuzzy primates who have their own life in the world, and unlike most others now or in history, are warm and safe this night.

My Prendick gaze has never left me. I see our obsession with gossip, our need for inclusion and comfort, our breakdowns under stress and when in pain, our bones articulating, our guts gurgling, our muscles flexing, our nerves firing, our oozy fluids secreting, our social exploration and cover-ups, our contemplations and sudden reactions, and our constant management of basics like food, shelter, sex, and family. I see your face: an ape face like every human being's, which is no insult or put-down, no affront to your humanity, which is after all a relictual riff on ape-dom, only as noble or as debased as you might assign to any living species, which is to say, neither.[2]

Alienation? No. Not when it becomes a deep sense of closeness, a sudden thought that every one of us fuzzy fellow creatures has his or her own developed and neural history, including a direct experience with pain, which becomes a desire to connect in ways that anyone and everyone could, but for some reason do not. Not when we look forward to the constructs of one another's mind that we might communicate, and the potential to combine with and reshape each, literally at the level of microglia and zones of the neocortex.

We are things—what kind of things? Living things, as with every other living thing. We are animals—what kind of animals? An ultrasocial great ape with its own species features, which we already know intimately. What does that mean? I don't think anyone knows that, but like Prendick, we refuse to look and spin into fear and fantasies when we're forced to. But one step further, in seeing ourselves as animals, I suggest it's neither a diminution nor the corresponding elevation—the Man/Beast divide is simply gone. One sees all of us-and-them differently, here in the world, with our histories as they are, our features as they are, and, in my case, keeping my mouth shut about it, until now.

Readings

It's heavy going, but neuroscience is best met through textbooks, such as Georg F. Striedter, *Principles of Brain Evolution* (2004); and Ann Butler and William Hodos, *Comparative Vertebrate Neuroanatomy: Evolution and Adaptation* (1996). Some of my points were informed by Randy J. Nelson and Brian C. Trainor, "Neural Mechanisms of Aggression," *Nature (Reviews)* 8: 536–546, 2007. You will find little or no mention of the triune brain in neuroscience. The term "limbic system" has sort of been retooled, mainly as a regional rather than functional descriptor; David Coleman has unfortunately reinforced the incorrect model with his term "amygdala switching."

Most of the sociobiological references in the previous chapters are relevant here, especially Chapter 7 in Jerome Barkow and Leda Cosmides (editors), *The Adapted Mind* (1995); and Matt Ridley's work.

A steady stream of anthropological and psychological books has continued the killer-ape or plains-ape concepts found in Desmond Morris's *The Naked Ape* (1967) in the 1960s. Most are startlingly simplistic, invoking stereotyped animal savagery, presenting violence as the core or essential feature of a given nonhuman profile, relying on the perceived role of other apes as protohuman, gender analysis that seems to use no measurable variables, as with Dale Peterson and Richard Wrangham, *Demonic Males* (1996); and David Livingstone Smith, *The Most Dangerous Animal* (2007). More philosophically inclined ideas are found in Roy F. Baumeister, *Evil: Inside Human Violence and Cruelty* (1999); Jack David Eller, *Violence and Culture* (2006); and David Churchman, *Why We Fight* (2013). A gentler or contrasting view is suggested by Dave Grossman, *On Killing* (1996); and Stephen Pinker, *The Better Angels of Our Nature* (2011); but the entire discourse is still separated enough from basic behavior and mammalogy that it seems ungrounded. A lot of the texts seem unable to parse differences among individual assault, mob violence, institutional violence, and war.

Fascinating values and category confusion can be found in Edward Payson Evans, *Animal Trials* (1906), which describes the extensive legal history of nonhuman animals charged with human crimes. It can be productively paired with Jason Hribal, *Fear of the Animal Planet* (2013), which presents a radical thesis about captive animal resistance.

Other analyses of *The Island of Doctor Moreau* focus on its sociological and symbolic content, as with Jennifer DeVere Brody's chapter "Deforming Island Races" in her *Impossible Purities* (2012). I think the novel requires extensive review regarding classicism and racism, as these issues are always found embedded solidly in individuals' viewpoints, with the plot as a whole typically tearing those viewpoints down. Moreau's and Montgomery's speech lapses into racism, as with "Kanaka" and identification of a Negroid type, but if two characters' worldviews were ever shown to be false by any story, this is the one. Useful as the current analyses might be, I think they potentially deflect the novel's potential to challenge these views.

The instances of lifeboat cannibalism are discussed in A. W. Brian Simpson, *Cannibalism and the Common Law* (1985); Neil Hanson, *The Custom of the Sea* (2001); and Andrew Walker, *Is Eating People Wrong?* (2010). In case you're interested, Brooks wasn't charged because the court needed a witness, which is bizarre since he had fully participated, and Dudley and Stephens were convicted of murder and sentenced to death, but given the judge's profound abuse of the court process, the sentencing was referred to a special royal court and they ultimately served six months imprisonment.

The "nay" side of the anthropological dispute over ritual cannibalism begins in 1979 with William Arens, *The Man-Eating Myth*, and more recently, Gananath Obeyesekere, *Cannibal Talk* (2005); the "yea" side is represented by Simon Mead and others in a widely publicized article in *Science* 300: 640–643, 2003, with a lot of popular media playing it up, and by Timothy Taylor, *The Buried Soul* (2005), but reanalysis of the relevant research has cast doubt on its conclusions.

Antonio Sanna's "Transforming Monsters into Humans" in *The Monstrous Identity of Humanity* (2007, Martin C. Bates, editor) offers the extreme and interesting idea that the entire island adventure is Prendick's deranged cover-up for having committed cannibalism in the lifeboat. I am not quite postmodern enough to buy the thesis as such, but the argument does capture the dynamics of characters' emotions.

Notes

1. A great deal of the academic intellectual tradition seems to have been completely convinced that nonhumans frequently engage in both cannibalism and incest, and that only humans have suppressed such urges through intensive cultural training. It's been a persistent idea perhaps due to familiarity with domesticated breeding, but behavioral ecology has shown it to be incorrect. Most of Freud's *Totem and Taboo* (1913) is dedicated to this fallacy, and Huxley briefly references incest in this context in *Evolution and Ethics*. Today's discourse is even more committed to the Man/Beast divide than theirs, so it has been difficult to address the ideas of shame, taboo, and morality when discussing a behavior that we did not invent and is apparently a thoroughly ordinary thing.
2. Racist human rhetoric frequently cites "monkey" or "ape" features in the targeted group, especially regarding the face. In itself, the observation is correct, but significantly, *always* correct. The racist's flaw isn't in seeing the ape in others' faces, but to deny it in his or her own, to insist that oneself is not in the Valley but that this other person is.

10

Big Thinks

The Ape Man can never be truly Man, can he? He gets so excited about big ideas and abstractions, he can't stop talking about them and wanting others to join him. He says this stuff is important; he wants it to be important even when it takes him into silly spirals. He's the talkative, geeky intellectual of the Beast Folk, and he probably annoys most of them even worse than he does Prendick, who never affords him an ounce of respect.

I have no claim to better mental processing than the Ape Man, and my conclusions may be as easily mocked or manipulated. But like him, all I can do is try.

Suffering and Philosophy

People have asked me, what's your book about? Technically, human exceptionalism. Historically, evolutionary biology and literature. But emotionally, I have found, it is about pain. Certainly and not trivially, the plot events of the novel are driven by the pain of bodily harm. This is most dramatically the pain inflicted on nonhuman animals during most of the history of live animal experimentation, and—as Moreau misses completely—it isn't solely about specific receptors in the skin, but rather the pain of helplessness, of fear, and of despair. Pain under torture is the issue. The immediate moral question is how much of this we intend to inflict on others. The bigger look is whether all life is helpless enough to be considered pain under torture. That may sound a little adolescent and melodramatic, but hold that thought for a moment.

Yet the novel also touches on the pain of being alone. When Prendick and the Hyena-Swine face one another on the beach, which I now see as the climax of the story, each is as alone as a person can be: sympathetically speaking, each is alienated from his culture, unable to regard its norms as a cosmic good; and also at his most repellent, each is potentially willing to be a cannibal. Their respective internal solitudes are the same, denied even the most basic metaphysical narrative, the one that says to you, "You are special."

What's worse, each of them hits every one of these buttons for the other. To each, the other is truly the Other—a presence whose perception and judgment cannot be controlled. Here, the terms from psychology, philosophy, and biology link up flawlessly: to our kind of ultrasocial animal, it is nearly intolerable that someone is looking at me in a way different from how I look at myself, and yet he or she has the temerity actually to be in the real world just the way I am. This moment is when we connect or we reject. In a world-experience lacking metaphysics, that becomes the only real question: when confronted by the Other, whether one feels exposed, isolated, foolish, targeted, and more alone than ever; or illuminated, given context, provided with listening, and no longer alone.

Reversion and Violence

If being a beast rather than a human meant no management over inflicting pain, no power to act against being inflicted with pain, uncontrolled aggression, randomly directed violence, predation arising from fury, eating of whatever presented itself (including one's fellows' bodies)—and if this were a beast's "place," its identity, such that elevating it to the rational state of Man could not help but fail—the "don't meddle" story would be so easy. It would look like this:

- Conversion into human anatomy and cognition (Man)
  - Management and responsibility of inflicting pain
- Reversion into the original animal forms (Beast)
  - Unmanaged aggression and violence
  - Predation
- Cannibalism

But in this story, these components don't lie down neatly into that dichotomized package. Instead, they are all present but curiously isolated from one another, overlapping in different ways and at different points. Their associations are cast into less comfortable forms: Moreau, in his idea of the rational Man, denying his responsibility to consider others' pain and therefore inflicting it to an unutterable degree; Montgomery, with his predilection for bad-tempered deadly force; Prendick, constantly fearful the Beast Folk will rend his flesh, while being the only character who really knows what it means to choose to eat another person.

Instead of the compartmentalized breakdown, you get an overlapping mess, which destroys the Beast + Violent + Cannibal/Man + Civil + Virtuous compartments. This isn't an interpretation or an impression or a symbol; confronting you with the mess is what the story does. Instead of "don't meddle," it's the "reality slap" story. Forget it, "Man." You aren't *what* you thought, you aren't *who* you thought, and in fact, it's not even *about* you. The real manager of violence in the story is M'Ling. The real moral voice regarding pain is the Puma Woman's.

A "don't meddle" story may have thrills and chills, but ultimately it's comforting. This one isn't. It shocks, it confronts, it prompts deflective interpretations that immediately deny the text, and it prompts revisions that alter its content into palatable forms and culturally silence it.

Why has this novel received such a strong recoil? Because if you experience the "reality slap," then you're no longer the hero of the Story of Man, whatever that might be for you. You're a Beast Man or Woman, standing in the Valley. There is no arc cresting in you or a shining goal you may point toward. Nothing in the universe or its processes protects you from crushing agony or death. All you have, and really nothing more, is that like any creature anywhere, you're at least undeniably here for a while—and after the shock, maybe that's enough. It's not *nothing*, literally. We're *something*, we're *someone*, grubby and brief as it might be, and we might think about working something out together, even if there's nothing so very grand about the task.

Questions

The story includes blank spots, apparent only because Prendick sees some hint of things going on outside his perception, or a late moment in a sequence of events. I'm not talking about the narrator of the story telling us an out-and-out lie. This isn't a puzzle or a guessing game. Instead, the narration accords strictly with Prendick's outlook and assumptions, but the events show how limited those are, therefore forcing the reader's participation in the story's content.

- Did the Puma Woman secretly weaken her chain in planning her escape?
- Did Prendick have sex with a Beast Woman?
- Did the Sayer of the Law act upon his oath to uphold the new Law when he killed Montgomery, whom he had just seen threaten Prendick?
- Was the Hyena-Swine sneaking up on Prendick on the beach, or trying to talk to him?

The "found document" technique isn't merely a quirk, but literally makes you, the reader, decide "yes" or "no" to these things. It's perhaps the most confrontational feature of the novel and, technically speaking, closes the Man/Beast divide with some force.

That brings the discussion straight to Arthur Schopenhauer, and why I decided to hook the sections of the book together with him. Believe me, I held back (you can mine this guy for quotes all day):

> Nothing leads more definitely to a recognition of the identity of the essential nature in animal and human phenomena than a study of zoology and anatomy. (Arthur Schopenhauer , "A Critique of Kant," *On the Basis of Morality*, p. 233)[1]

I don't think Schopenhauer's writings are a recognized school of thought with a codified set of precepts, and even if they were, I'm not interested in the traditional philosophical model of a worldview or a construct to adhere to. Instead, Schopenhauer interests me because he breaks with—and in fact completely leaves behind—the idea of cosmology as a moral directive. The cosmos provides no spiritual or logical guidance, and instead is imbued with the Will—the unexplained and quite unstoppable, defining drive of doing things. Today, I'd talk about this as a feature of thermodynamics, but that doesn't really matter—the idea is that natural forces are amoral and overwhelming, and our own behaviors, drives, and perspectives are local examples.

That's an important point, that Schopenhauer's Will is not merely a fun little celebration of human desire. It's not about saying, "I want a pony," and making that happen. It means that any drive, any sense of achievement or purpose—family love, community or national identity, a political end—is as amoral and potentially full of destruction and suffering as anything else. It's ultimately saying that our sincere sensation of purpose is not the same as having purpose. It's not discoverable through logic, through investigation of the physical world, or through an ideal. Really, you're just "here."

In practice, to the extent that ethical suggestions are even present, these notions fall back on simple daily compassion and trying to muddle through the day without letting your drives sucker you into cruelty or misery, with no concern for a grand design to bask in or to seek.

This trajectory of thought didn't connect with scientific views very much at the time, which is why seeing Schopenhauer so explicitly in *Evolution and Ethics* is a bit surprising, especially since he's uncredited in what's otherwise a very well-referenced work. One might do worse than to trace intellectual consistencies from David Hume's writings through various pre- or proto-evolutionary thinkers like La Mettrie, then Lawrence of course, through Schopenhauer's body of work, pulling apart the *Vestiges* to see which ways it zigged and zagged, through Huxley's and Darwin's extensive work on anatomy and behavior, Huxley's biography of Hume, Nietzsche's *The Gay Science* among others, and, via *Evolution and Ethics*, finally into the renewed debates about human behavioral ecology beginning in the 1980s—particularly how badly they've floundered.

Such a project would uncover a surprising idea about human ethics: that I might prefer to be an "earthy" animal, in Nietzsche's terms, subject to physicality and history, in the absence of natural law, who can at least try to arrive at less suffering among us, than a higher Man. That higher Man hasn't done very well for us; he seems defined by his empowerment to inflict suffering or, upon being demoted, forced to receive it at another's whim. Maybe after all the pretty talk is past, that's what the higher Man really is: the Master of the House of Pain.

Science and Humanity

What happened to philosophy might be known to other historians and scholars, but it's a mystery to me. All I know is that after about 1900, the discipline's prior connection with natural history evaporated. It has a terrible relationship with nature, which stubbornly refuses to be either a deontological principle or a utilitarian end, and, as far as I can tell, seems to focus strictly on either metaphysics or a disembodied and over-rationalized form of ethics. The twentieth-century extension of Kierkegaard and others, existentialism, went off on its own road with a spiritual notion of personal free will that maintained nothing of Schopenhauer's concept.

I could be missing the boat regarding the most famous philosophers of science, Karl Popper and C. P. Snow, but with all respect to them, their works are over-concerned with epistemology and not enough with content. I don't see that they had much direct contact with science in action or, in Snow's case, at least not much nuts and bolts biology.

As I've encountered it, the philosophical (and eventually others') rejection of science is based on a series of apparently deliberate mistakes, in a specific order.

- First, that reductive causes, or small things making big ones happen, are considered more important than—here a word is lacking—the reverse, with big things making small ones happen. Biologists study both and always have.
- Second, reductive cause is confounded with, or smeared as, reduction-*ism*, the idea that larger-scale phenomena are deprived of their identity or scale-specific properties if their reductive causes are identified.
- Third, reductionism is confounded with atomism, the idea that a phenomenon's features can be identified in full in each of its tiny subunits (genes are particularly vulnerable to this misunderstanding).
- Finally, and most toxic of all, plain and simple material thinking is identified with this weird construct of reductionism + atomism. Therefore to call for a scientific understanding of humans is instantly derided as essentialist, the claim that one is reducing the identity of humans into mindless little particles. It's not much different from Abernethy's outrage at Lawrence, merely given an intellectual gloss if you don't mind the errors.

Scientific inquiry is guilty of none of these things, but it did itself no favor internally either. The new social environment of science is riddled with awful, senseless dichotomies, of which the mindless/brainless distinction is a mere echo. These dichotomies—instinct/learning, genetic/environment, nature/culture—are present in force throughout science and non-science, tripping up every effort to link across those disciplines. They're perpetuated and reinforced by obfuscating terminology, maintaining the Man/Beast dichotomy at every turn, such that complex and/or social behavior in nonhumans is "instinct" even when it includes learning, memory, and emotions, whereas even simple behavior in humans is "conscious."

The reluctance has always remained in the bedrock of modern biology, systematics. Linnaeus's thoughts on humans and the other apes were confirmed by Huxley in the 1860s, but the muted, misleading classification was never revised to reflect his work. In George Gaylord Simpson's great taxonomic revision of mammals in 1945, there it was again: *Homo sapiens* by itself, way over in a family called Anthropodidae, with the other great apes racked together against all anatomical logic in Simidae. When data from mitochondrial DNA finally made this false construct too absurd to ignore in the 1980s, it was wrongly billed, even by biologists, as overturning earlier anatomical work, when it actually confirmed and refined what Huxley had shown over a century before.

That was no exception, far from it. It's still like pulling teeth to acknowledge that the orangutan, gorilla, chimpanzee, and human, along with their extinct related species, comprise a single taxonomic family, let alone to debate its name or to address the unnecessary over-nominalizing within it into tribes and super-families and subfamilies.

I could kick some shins regarding professionalism in the sciences, too, in how what began as an unconstructed set of grassroots journals become a starchy, risk-aversive, and cloistered activity subordinate to publications, reputations, promotion, and tenure, and effective servitude to university administrations through the financial mechanisms of grants. A certain tension has always existed in scientific education between the implications of (i) professional skill and employment and (ii) specialization and segregation from other branches of knowledge, and it's long overdue for some historical scrutiny.

Academe as a whole went ahead and firewalled the whole damn thing from ourselves, as well, mainly through disciplinary boundaries: psychologists get the mind, physical anthropologists get the body form and evolution, cultural and social anthropologists get the habits and actions, biologists (or rather, medical physiologists) get the diseases and pathologies, historians get the documentation. These boundaries are not minor—they define job success and the unspoken standards of content for scientific publishing, and quickly became perceived as intrinsic; to be a scientist at all, one had to internalize the standards within each one, separately.

I'm sad to say that both modern science and academia in general are practically defined by coping mechanisms to avoid being scientific: all those trappings of shiny science, safe in its segregated fancy buildings; all those nice "-sciences" appended to the department names; and little to none of the thinking that throws open doors across the established disciplines, prompts the sharpest of personal and political reflection, and holds every claim made by anyone to the fire of making sense. We see no Lawrences today—there's a dedicated social system in place to make sure of it.

By the mid-twentieth century, real scientific inquiry wouldn't touch humanity with a ten-foot pole. Dialogue is belittled by appeals to consciousness, to metaphysics, to recent details like industrialization, to the dubious honor of overpopulation, to flat-out untruths, and to disciplinary constructs and constraints. It's especially disheartening to encounter how badly biology has been demonized in the social sciences as ethically and politically bankrupt. I see there an outright flat refusal of psychology, sociology, economics, and a significant component of anthropology to make Lawrence's and Darwin's jump. Instead, there's a sprayed fog of claims that we are "unlike the animals," or deceptively, "unlike the other animals," and "from the animals but" followed by a false claim. I see hardly a shred of thought toward economics as a subset of ecology, sociology as a subset of behavioral ecology, and psychology as a subset of developmental psychobiology. Instead, those are supposed to be something else, the latest word for that separate high status for humans: culture.

Ah, Culture

The word from sociology is that biology denies culture. I don't know where they get that, but there it is. Let's see what happens when I look at human culture with Prendick's gaze, but unlike him, without fear. What happens, how terrible is it, to look at the human species just as one looks at a species of gopher?

- It has a phylogenetic and species identity, expressed as a distinct set of developmental events.
- It has an ecological context composed of abilities and constraints.
- It has an identifiable range of behavior and socializing.
- It has regional variants regarding fairly trivial details and a considerable amount of gene flow.

For any and every species, the biologist expects these topics to be integrated and coherent, with acceptable results being in part defined by consistency across different analyses. He or she knows that any of these topics might boom into a rich and nuanced array of history and surprising properties. We're supposed to debate how they work in order to enrich, expand, and revise all possible connections and understanding.

Here, I see a relict species of bipedal ape with its own profile of reproductive, cognitive, and social features; specifically, it records what it communicates—and look, look at what this ape is doing! Briefly, these creatures communicate with abstractions, and they record their communications—then, the recordings become a learning environment for a new generation. The information, values, standards, and vocabulary for a given location therefore become a distinctive spin, reflecting the history there. So, I see culture, always proliferating and changing, always arriving as a distortion of the past and always affected unexpectedly by what happens next. All right, that's not controversial, it's an observation. No one is going to take that away.

Here's how it works (Figure 10.1). The biological animal in a given place has all the usual features: a life-history strategy, a set of behavioral parameters,

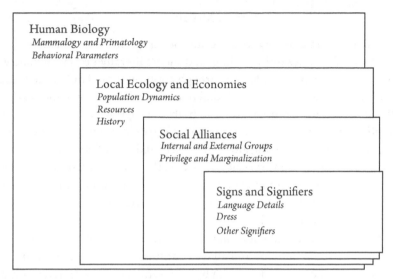

Figure 10.1 Human cultural biology or biological culture, as you prefer.

a population size, a resource base with specific geographic properties, and any number of details like local parasites. Given its abilities with social organizing and symbolic communication, various subcultures of effort, exchange, and recompense are operating there, however contentiously and with however much pain and exploitation; these operations and their history are absolutely the historical products of the creature's biology, including cognition, in this place. These social arrangements become themselves an operating environment, learned and experienced as rock-solid identities and tensions, reinforced through education, but also subject to revision and upheaval as conditions in the underlying power and resources change. These identities and tensions are made explicit through local terms and symbols, in every means of expression.

It's clear, sensible, and easy. I haven't dented or devalued anything concrete stated by any of these other disciplines, I haven't said they're "nothing but" biology, and I haven't said that we as humans don't do those things. The only thing that's added is to expect that these different levels make sense relative to one another, such that no set of physical rules gets to be magically isolated from any others. The only things now missing are presumed elevated functionality, special history composed of heroic narrative elements, intrinsically admirable achievements, and a presumed destiny.

The past couple of centuries have brought some new details: industrial technology and notable population abundance, currently undergoing exponential growth. All right, that can be a new topic, and maybe this construct can help us change dilemmas into questions.

Morals

I have no intention of trying to argue anyone out of the Man/Beast divide—it is a very deeply felt, unexamined thing, and there's literally no way anyone will be dislodged from his or her side of it through confrontation. I'm here to talk about what it's like to see no categorical difference between humans and nonhumans at all, using the term "animal" to mean both, and significantly, seeing absolutely no implications concerning ethical or social capabilities by using that word. My criticism is that *this* viewpoint, although present and identifiable in individuals like myself, has no general cultural voice. It died on the vine right around 1900, even in the sciences, and therefore has failed to develop any philosophical, ethical, political, or artistic identity for over a century. We need to recognize that failure, to identify its origins, and effectively, to call it *out*.

Specifically: if you find yourself, like me, unable to identify any fundamental distinction between humans and all other animals, and if you are similarly baffled by the implications so deeply felt by others, then the question for us must be, "What ethics do you stand by, and where do they come from?" And again less formally, "Well, then, who *do* you think you are?" Furthermore, these questions are not merely individual curiosities, but must be extended outward, socially, intellectually, and politically, to their applications in society, whatever those may be.

Huxley thought he knew, or at least he phrased it optimistically in the last few minutes of his talk. Scientific context and new ethical thinking would bring us around, to solve it somehow or to save us, in the 1890s or by a century later, to blossom into a way actually to live among one another. He was praising the gains in understanding in astronomy, physics, and chemistry, using them as a model for what these other disciplines may become. "Physiology, Psychology, Ethics, Political Science must submit to the same ordeal" (*Evolution and Ethics*, p. 84 in original; p. 142 in Princeton University Press edn.). This was his key to arriving at a more just society, his "great work of helping one another" in the constant negative context of a grim history and array of easily tapped grim behaviors.

Boy, was he ever wrong on that one. Astronomy, physics, and chemistry arrived all right, transforming the world into a war ecology and war economy, with a global impact on ecology best described as disastrous. The disciplines he hoped to see develop have become methods of political control beyond anyone's imaginings.

Was Huxley even theoretically right? One would have to admit to a century of dedicated "not quite there yet," an unstopping reiteration of the 1890s and 19-oughts—the same peripheral imperial assault on the central empire of Eurasia, the same extraction from all economies to support it, the framework of the Boer War, the opium trade, the destruction of the Ottoman Empire, all still in place, repeated in new places, and elaborated upon with more propaganda, more atrocity, and more cumulative misery, like a horrific Mandelbrot's snowflake. I may be a pessimist, but I'm inclined to say that Huxley's future ethical developments, the ones he said would be "irrational to doubt," aren't coming.

Is there a morality to discover? Can some perspective on the biological human, however battered and dismembered, be recovered enough even to see whether it helps us, as he put it, help one another? Maybe, if by "morality" we mean common ground for discussion, a context rather than a directive. Maybe, if we can keep biology from becoming a cosmology, a story in which we hold the starring role. Maybe, if we can grasp that basic ethics does not translate easily into sustained policy. Maybe, if the idea is more like building a bridge in the knowledge of Newtonian physics, and understanding that we might not be very good at it, and it's a work in progress. Maybe, if we don't succumb to the Naturalistic Fallacy of saying "what's biological is right" or—in the lack of a name for it—the corresponding Humanistic Fallacy of claiming culture and policy exist free-floating from our biology. And finally, maybe if we can decide that nothing about morality or an ethos is about making anyone happy, but about reducing misery, that it works not because everyone is satisfied but because we have some agreed-upon idea of what shows we're better off than we were.

And after that, I want a pony.

Science and Fiction

All that is too heady for me, even in Ape Man mode. This was a book about a book, and staying down in that scope, it turns out that over the years, science fiction—or speculative fiction—has plenty to offer in its confrontational mode. When I was in college in the 1980s, professors reacted to the suggestion that "genre fiction" had something to offer with contempt, even nausea. Times have changed, though, and some room has been made after all. Two references that could form the basis for a whole new subdiscipline along these lines include Jonathan Gottschall's *The Storytelling Animal* (2012) and Sherryl Vint's *Animal Alterity* (2012), as well as her paper "Animals and Animality from the Island of Moreau to the Uplift Universe," in the *Yearbook of English Studies* (2007).

I'm talking about reading science fiction not as prediction but rather as contemporary insight, dated exactly at the moment of its writing. Its trappings and excess aren't escapism, or not entirely, but rather act as genuine confrontation. I think the whole array of science fiction, fantasy, and horror can express and develop sensitive real-life topics in ways that naturalistic stories and essays cannot.

I'd even suggest staying away from the most well-known or most-lauded authors, whose work tends to be more confirmatory, such as Arthur C. Clarke, Robert Heinlein, and Isaac Asimov, although isolated works can qualify, as with the latter's *I Robot* stories. It's the dissenting, weird ones I recommend: Fredric Brown, Jack Vance, Alfred Bester, and especially Fritz Leiber; and from the freewheeling 1960s and 1970s, wilder work from Harlan Ellison and his editorial work

Dangerous Visions, Norman Spinrad, Philip K. Dick, James Tiptree Jr. (Alice Sheldon), Stanislaw Lem, Arkadi and Boris Strugatski, and Gene Wolfe, just to name a few. The engineering-style, technical branding of science fiction during the 1940s and 1950s seems now to me like a minor editorial artifact, as the more psychedelic and less concerned with technical accuracy it became, the harder the "reality slap" it delivered.

Although "reality-slap" science fiction is not as prevalent in cinema and TV, the minority thereof is strikingly powerful. The original *Star Trek* series is almost entirely submerged beneath the weight of the franchise and fandom, but some day perhaps it could be revisited as itself, especially in partnership with the British show—almost an extension of it—*Blake's 7*. The first four *Planet of the Apes* films may represent the high-water mark of questioning human exceptionalism in any endeavor whatsoever for the twentieth century. In series media, sometimes the tension with "don't meddle" content is a productive thing of its own, and these all provide a lot of content for that analysis—maybe more than any number of more formal, more academically recognized discussions.

Well, that's it for the "Big Thinks": "read some science fiction." It seems a feeble thing, but where formal philosophy and science have fallen short, maybe *The Island of Doctor Moreau* still shows the way.

Note

1. Arthur Schopenhauer, *On the Basis of Morality*, 1818, included in *Philosophical Writings*, edited by Wolfgang Schirmacher, published by The Continuum Publishing Company, 1994.

GLOSSARY

All of the following terms are described according to my understanding and in my own phrasing, and any limitations or errors, if present, are my responsibility.

Absurd, The: The perceived lack of meaning or purpose in the course of events, life experiences, and morality; associated with existential philosophy.

Adaptive radiation: A model or perceived pattern of living organisms' speciation, in which diversity among types of creatures is thought to arise from local selective (adaptive) factors in populations that have become geographically isolated from other members of the original species.

Agnostic: As originally coined by Thomas Huxley, the view that considers divine or other nonphysical purposes or influences to be irrelevant to life and morality, whether in terms of existence or in terms of effects, today called "hard agnosticism"; often misconstrued as "undecided" or "in between" regarding the existence of God.

Anglicanism: The branch of Protestantism originating in the breakaway of the Church of England from the Roman Catholic Church in the late sixteenth century, including the Churches of Scotland and Ireland as well; in the nineteenth century, the state church of Great Britain and its associated political, educational, and media connections; now a more generalized term, "Anglican Communion," meaning association with the Church of England but not necessarily in its authority, as with the Episcopal Church in the United States.

Animal: An organism composed of multiple cloned protozoa-like cells, specifically without cell walls and without the ability to synthesize sugar; all such organisms are grouped into the larger taxon Animalia.

Australopithecus: A genus of mammals, within the Primates, including several extinct species; as one group of bipedal apes, closely related to *Homo* and *Paranthropus*.

Behavioral ecology: A research discipline concerning the options, decision-making priorities, and life-history strategies of organisms in their ordinary environment; a combination of the older discipline of ethology with debates concerning selection during the 1970s and 1980s.

Cell: the structural subunit of all living things observed to date, characterized by a membrane, metabolism, and mechanisms of reproduction based on a narrow range of chemical operations; technically, individual cells are the only literally living things.

Cell theory: A body of phrases and ideas concerning cells developed throughout the nineteenth century, considered a founding concept for modern biology; technically, life is cellular, a cell is the smallest unit of life, and cells are derived from prior cells; significantly, embedded in the intellectual position that living structures and processes are material phenomena, specifically a subset of chemistry.

Cerebral cortex/neocortex: Layers of neural tissue developed from the vertebrate forebrain (telencephalon); in mammals, it includes the neocortex, characterized by six distinctive layers; the neocortex is disproportionately large in primates, and for primates, disproportionately large in the genus *Homo*.

Chartism: A broad, decentralized British social movement beginning in the 1830s, associated with a wide range of dissent and reform, especially voter reform and rescinding the property ownership status required for Parliament seats.

Classification (Linnaean): The naming and arrangement of the known types of organisms into hierarchical, nested categories, using a system of grouping proposed by Carl Linnaeus, such that our species of human is a type of primate among others, all primates are a type of mammal among others, all mammals are a type of vertebrate, and so on; ultimately, all living things are placed into a very limited number of Kingdoms and Domains.

Cognition; cognitive map: The mental and experiential processing of perceptions into a working model of the world and one's position in it (identity in psychological terms), especially in terms of quantities, relationships, and conditional predictions; the parameters and sophistication of the map is considered to be highly species-specific.

Cognitive dissonance: The distress experienced when the borders and content of one's mental categories are challenged by direct evidence, usually resulting in defensive measures to reject the evidence, including compartmentalization, denial, and reactive anger.

Consciousness: This term and its near synonyms (sapience, sentience, awareness) may be defined in non-material terms, defined circularly only in reference to one another, are undefined, or when phrased in material terms, are applicable to all known living things; often confounded with the properties of the cognitive map as experienced by humans, or with the characteristic emotions and actions of humans.

Creationism: The political effort to maintain divine causes as explanations in education about the natural world, often phrased in terms of direct opposition to evolutionary theory or to a distorted version of it; in the nineteenth century in Great Britain, associated with the Anglican Church, and beginning in the late nineteenth century in the United States and continuing through the present, associated with evangelical Christianity in a number of coalitions with sectors of mainstream Protestantism; roughly divided among Biblical Creationism, so-called Scientific Creationism, and since the mid-1980s, Intelligent Design.

Density-dependent selection: A condition of circumstances in a given organism in which the strength and direction of selection are affected by the current prevalence of traits in the population of creatures, for example, when one of several is selected for only when it is rare; often results in a diversity of traits rather than a single most prevalent one.

Developmental cascade: The sequential series of genetic, chemical, and biochemical events by which a given body part is formed and becomes operational; strongly tied to the concept that single genes are almost never "for" a single identifiable body part or function, but instead that a gene's effect must be assessed in the context of many others, and that even a dramatic change in a single gene may be either invisible in the cascade's current effects, or may have different effects depending on the cascade it's in; a central concept in Evo Devo.

Differentiation: The changes in physical structure and chemical activity within the stem cells and semi-stem cells that compose a developing organ or body part, collectively resulting in distinctive tissue functions; follows morphogenesis.

Euthanasia: Killing in the context of ending pain and suffering; as applied to humans, controversial in terms of prior assent and assisted suicide; as applied to nonhumans, controversial in terms of research use and zoo animals' longevity.

Evo Devo: A slang term in biology from the 1990s referring to research concerning evolutionary processes, biological concepts of homology, species diversity, and the genetic, chemical, and biomechanical features of embryonic development; associated with many new research results concerning the similarity of developmental genetics across very different organisms, as well as the chemical properties and effects of morphogens.

Evolution (nineteenth century): Technically, changes in form and function of living creatures, resulting in a diversity of types and ways of life; the term underwent a number of uses and associations, notably Jean-Baptiste Lamarck's concept of progressive but individually generated change with Republican, that is, revolutionary or radical-reform political associations; a later association with an orderly or "proper" system of progress in the changes of living organisms; often misconstrued as a specific position on metaphysical or religious topics; often confounded with Charles Darwin's theory of natural selection, but its scientific relationship to his term "descent with modification," and his occasional use of the term "evolution" are topics of scholarly debate.

Evolution (modern): Changes in the form and function of living creatures in the context of material thinking, due to any physical causes, regardless of outcomes, without reference to progress, and without implications concerning purposes or appropriateness.

Evolution as entropy: As coined by Daniel Brooks, proposed integration of evolutionary theory and principles of thermodynamics, strongly emphasizing the role of phylogenetic history on living systems; here cited as an alternative view of patterns of speciation, contrasting with the causal details of adaptive radiation, in that speciation is interpreted as a likely outcome due to a variety of conditions and effects, with selection as a secondary rather than primary cause.

Existentialism: A body of philosophical ideas beginning in the early nineteenth century based on the lack of higher meaning (the Absurd) in existence and actions, specifically the absence of a thinking creative or motive divinity, sometimes strongly emphasizing human free will and the need for a new and humanocentric moral system; the works typically identified as such were not necessarily written to be consistent, and the "movement" or group was not named or semi-codified until the 1920s. Earlier authors include Arthur Schopenhauer, Søren Kierkegaard, and Friedrich Nietzsche.

Fitness: In the theory of selection, the standing of a particular variant in terms of reproductive success relative to other variants at a given historical point of assessment; it concerns only the historical reasons for the current prevalence of the variant and lacks content or implications regarding the rightness or goodness of the variant, or of its implications for the entire species; its use in Spencer's phrase "survival of the fittest," which refers either to whole individuals or whole species, depending on the interpretation, has led to perhaps the widest degree of incorrect and negative misuse relative to a scientific theory of any term in history; also widely misunderstood to imply a matching or "fit" between organism and environment, which is better understood as a topic of debate within biology.

Future shock: The psychological and social condition of experiencing life-altering changes in technology too fast for generational and cultural adjustments to transform them into comfortable or easily processed moral constructs; coined by Alvin Toffler in the book of the same title (1970).

Glucose: One of the two simplest carbohydrates, present in body fluids either through ingestion or through the processing of glycogen in the liver; the direct source of external energy for almost all known living cells.

Great Awakening(s) (First, Second, Third): Specific historical increases in the membership and diversity of Protestant sects, primarily in the United States but also in England and other nations, typically associated with periods of political mobilization and controversy.

Homeostasis: The physiological effect of stabilized conditions within a cell or within a given area of the body, whether of water content, salt concentration, levels of hormones, temperature, or other variables that affect living functions.

Homo: The genus of mammals, within the Primates, which includes the living species *Homo sapiens* and at current count six or seven extinct species; characterized by a large neocortex and correspondingly bulbous cranium. With the extinct genera *Paranthropus* and *Australopithecus*, often grouped as the Homininae or Hominidae, the bipedal apes.

Institutional Animal Care and Use Committee (IACUC): The individuals required by US law to assess the treatment of animals in research and other use, composed of a rotating membership among the institution's staff and required to include a veterinarian and at least one outside member; a given project or procedure may not be implemented without IACUC approval and continued approval through periodic review.

Intelligence: This term is not broadly defined in biology and is considered problematic when compared across species; when used, it applies to an identifiable performance variable within a given species' cognitive range, and does not correspond to the conversational meanings of "smart" or its converse, "stupid"; its use toward humans, especially in reference to academic grades and so-called intelligence testing, is generally not highly regarded as a meaningful variable.

Life-history strategy: The timing and amounts of energy expended by a living creature throughout its life, described at the level of an entire species in terms of individual alternatives and options; typically the term lists longevity and its components, such as time to maturation, the number of offspring produced both overall and in given instances, body size at different stages of life, modes of reproduction such as internal versus external gestation, modes of acquiring and storing energy, and specific strategies of energy use such as metamorphosis, hibernation, or season-specific mate competition.

Macromolecule: A group of molecules based on carbon, oxygen, and hydrogen, with the property of increased size and complexity with added subunits; includes carbohydrates, lipids, proteins, and nucleotides; the primary features and functions of living cells derive from these molecules' interactions.

Material(-ism): In the context of nineteenth-century scientific controversy, the presumption that living processes are a subset of chemistry and the larger category of physics; unrelated to the later implications of a values focus on wealth or possessions.

Metabolism: The transfer of energy holding glucose together into energy holding adenosine triphosphate (ATP) together, within a cell; ATP is the within-cell provider of energy for all cellular activities. In biochemical terms, metabolism is the defining feature of life for almost all known living cells; contrary to popular presentations, it does not violate the second law of thermodynamics.

Morphogenesis: The arrangement of stem and near-stem cells into the shape of an embryonic organ or body part, through a combination of growth, apoptosis (cell death), and cell death; strongly affected by the presence, absence, and concentration of a family of chemicals called morphogens; precedes differentiation.

Neurohormone: One of several signaling chemicals that is transported along neural tissue from its origin to a target organ or area of an organ; distinguished from hormone (released into blood) and from neurotransmitter (at the junction of nerves), although the same chemical sometimes acts in two or three of these capacities.

Other, The: The recognition that another person is observing oneself in a fashion that one cannot control or with results that one would prefer not to acknowledge, resulting in a violation of the narrative one builds about his or her identity; originally associated with existential philosophy and now more widely applied.

Oxford movement: An influential political effort during the mid-nineteenth century in England to strengthen and standardize the Anglican Church to be more like Roman Catholicism, and to align it with a variety of policy positions; criticized by its opponents as authoritarian, privileged, and anti-intellectual, as encompassed in the term "Establishment."

Pain categories: The IACUC rating of animals' experience in research and other scientific use, graded from A, which is never applied as it presumes complete lack of stress or pain due to human use, to E, which indicates unalleviated pain and/or death due to the experimental conditions; criteria for the assigned category are expected to conform with research results concerning animals' own behavioral and physiological indicators of pain and stress.

Paranthropus: A genus of mammals, within the Primates, including several extinct species; as one group of bipedal apes, closely related to *Homo* and *Australopithecus*; characterized by large and muscular jaws and correspondingly large flat teeth.

Pleistocene: A geological epoch lasting from about two and a half million years ago to about 118 thousand years ago, characterized by repeated waves of continental glaciers and also by a high rate of extinction for large mammals. The genus *Homo* evolved during this period.

Religiosity: The personal experience or perceived experience of contact with divine, spiritual, or similar extra-physical and purposeful forces; in this text, distinguished from observance, community identity, institutions, and doctrine.

Selection, natural (original phrasing): Contrasted with sexual selection, in which the selected features concern mating success, and with artificial selection, in which the reproductive options, and therefore the selected features, are under human control; often misconstrued to apply to species' success in defeating other species in a "war of all against all," a fictional or poetic construct.

Selection (modern phrasing): The process by which traits' prevalence within a species reflect reproductive history, in terms of which traits have afforded the most reproductive success in the past; it applies only to traits with significant heritable components; in this construction, the original constructions of artificial, natural, and sexual selection differ only in the details of which aspects of life and environment happen to have affected reproductive success most strongly.

Social Darwinism: During the nineteenth century, a little-used term employed most frequently in relation to progressive activists who cited Darwin's work or some paraphrase of it; during the twentieth century, defined and popularized by William Hofstadter as the industrial and imperial paraphrase and distortion of Darwin's ideas, especially Spencer's phrase "survival of the fittest," as retroactively exemplified by Andrew Carnegie and John D. Rockefeller, and also associated with Nazism by Hofstadter.

Sociobiology: Technically, the topic of evolved social behavior in the context of behavioral ecology; often narrowly identified with the interpretation of human behavior in that context; expanded or redirected into human-specific studies in the form of evolutionary psychology during the 1990s.

Speciation: The appearance of a recognizably new species through modification of existing organisms, occurring through a variety of processes including selection and genetic drift, but in almost all cases occurring cladistically, meaning in a subdivision or small group of the prior species, as opposed to a blanket modification of the prior species as a whole (anagenesis); it is a macroevolutionary phenomenon, often confounded with the microevolutionary phenomenon of selection.

Speculative fiction: Harlan Ellison's term to describe imaginative or fantastic fiction that challenges social mores and familiar parameters of identity; coined to challenge the commercial terms "science fiction" and "sci-fi," to eliminate the artificial division between science fiction and fantasy, and to remove the former term's focus on technology and futurism as a defining feature.

Survival of the fittest: Herbert Spencer's phrase to summarize Charles Darwin's theory of natural selection; included in the sixth edition of *The Origin of Species* and widely misattributed to Darwin, also widely interpreted to identify Darwin's ideas with ruthlessness, betrayal, greed, exploitation, racism, and imperialism.

Systematics: The combination of taxonomy with phylogeny, such that the naming of organisms conforms with the evolutionary and biogeographic understanding of their historical origins.

Taxonomy; taxon: The general term for a named type of organisms or a group of such types, in the context of Carl Linnaeus's system of classification; the finest formal grain of taxonomy is the designation of a species, and the most general, or "biggest box," was originally called Kingdom, and more recently has been expanded to an even larger category called Domain.

Ultrasocial: The social category in which one's position, relationships, and reputation may well exceed other limiting factors on reproductive success, and in which a great deal of one's cognitive map concerns current social categories and nuances; highly cooperative but also potentially full of aggression, exploitation, and deception; sometimes applied strictly to humans, but sometimes more generally to a variety of mammalian, avian, and other species; contrasted sharply with eusocial behavior, observed in a number of insects and a few other species, in which highly cooperative behavior is displayed by closely related individuals, such that the community is effectively a tightly knit family, and in which conflicts of interest are rare to absent.

Victorian Age: Technically, the period of Queen Victoria I of England, from June 1837 to January 1901, closely related to but not synonymous with the British Empire, which dates from 1815 to various identifiable points in the early to mid-twentieth century; more generally, referring to British global imperial policies and wars, the immediate consequences of the Industrial Revolution, a distinctive profile of domestic reforms and oppressions, a distinctive set of values elevating Anglo-Saxon ethnicity and outward prudishness, the establishing of English as the world's most widely spoken language, and a host of other positive, negative, and ambiguous associations.

Vitalism: Originally, a reference to "vivifying" forces that are not derived from chemical or other physical phenomena and that impart the quality of life to matter; sometimes more broadly applied as spiritual or metaphysical properties; in all cases, in reference to humans, associated with the capacity for moral functions and a potential awareness of divine or spiritual purpose.

Vivisection: A pejorative term for interventional research using animals, highlighting the suffering either actually occurring or believed to occur due to this usage.

Will, The: In Arthur Schopenhauer's writings, the experience of living beings as expressions of driving, directionless, appetitive, and amoral natural forces; the degree to which this is a metaphysical rather than physical concept is a matter for debate.

INDEX